From Man to Ape

From Man to Ape

Darwinism in Argentina, 1870–1920

ADRIANA NOVOA AND
ALEX LEVINE

THE UNIVERSITY OF CHICAGO PRESS   CHICAGO AND LONDON

ADRIANA NOVOA is assistant professor in the Department of the Humanities and Cultural Studies and ALEX LEVINE is associate professor in the Department of Philosophy, both at the University of South Florida.

The University of Chicago Press, Chicago 60637
The University of Chicago Press, Ltd., London
© 2010 by The University of Chicago
All rights reserved. Published 2010
Printed in the United States of America

18 17 16 15 14 13 12 11 10   1 2 3 4 5

ISBN-13: 978-0-226-59616-7 (cloth)
ISBN-10: 0-226-59616-8 (cloth)

Library of Congress Cataloging-in-Publication Data
Novoa, Adriana, 1963–
  From man to ape : Darwinism in Argentina, 1870–1920 / Adriana Novoa and Alex Levine.
     p. cm.
  Includes bibliographical references and index.
  ISBN-13: 978-0-226-59616-7 (cloth : alk. paper)
  ISBN-10: 0-226-59616-8 (cloth : alk. paper) 1. Evolution (Biology)—Argentina—History.
2. Darwin, Charles, 1809–1882—Influence. 3. Science—Argentina—History. I. Levine, Alex, 1966– II. Title.
  QH361.N68 2010
  576.8'20982—dc22
                                                                                        2010015031

♾ The paper used in this publication meets the minimum requirements of the American National Standard for Information Sciences—Permanence of Paper for Printed Library Materials, ANSI Z39.48-1992.

FOR MAIA

Contents

Acknowledgments ix

Introduction 1

PART I

CHAPTER 1. The Roots of Evolutionary Thought in Argentina 17

CHAPTER 2. The Reception of Darwinism in Argentina 51

CHAPTER 3. The Triumph of Darwinism in Argentina 83

PART II

CHAPTER 4. The Culture of Extinction 119

CHAPTER 5. Sexual Selection and the Politics of Mating 156

CHAPTER 6. Evolutionary Psychology and Its Analogies 192

Conclusion 228

Notes 239

Works Cited 251

Index 275

Acknowledgments

This book has been a lengthy and challenging undertaking for both of us, and it would never have reached completion without the aid and encouragement of family, friends, and colleagues too numerous to mention. We hope that those whose names we omit will forgive us.

The original idea for this book arose over the course of discussions between a historian of modern Latin America (Novoa) and a philosopher of science (Levine) during the late 1990s. In 2004 it was nourished by our colleagues in the Philosophy Faculty Seminar at Lehigh University, to whom we presented early drafts of critical arguments and translations of some of our key sources. We are long overdue in thanking Gordon C. F. Bearn, Mark Bickhard, Robin Dillon, Steve Goldman, Michael Mendelson, and Roslyn Weiss for their participation in this seminar, which lent us needed impetus. Also at Lehigh University, we benefited greatly from the assistance of marvelous librarians, including Philip Metzger (Special Collections) and Pat Ward (Interlibrary Loan Wizard). Their willingness and ability to go the extra mile in obtaining physical copies of rare sources, published a long time ago on cheap wood-pulp paper, in various languages, and now in extremely delicate condition, cannot be sufficiently acknowledged.

At the University of South Florida, both of us have been the fortunate beneficiaries of Humanities Institute Summer Fellowships, which freed our summers for much of the heavy lifting. One of us (Novoa) also received a creative scholarship grant that allowed us to begin writing. Among our colleagues, we especially thank Roger Ariew, whose moral and material support, together with his considerable practical advice, led directly to the completion of this book. We also wish to thank Pablo Brescia and Joanne Waugh, in particular. Various aspects of this book have also made their way into our classes, as well as other interactions with undergraduate and graduate students at USF, to whom we are also grateful. The preparation of the bibliography was made immeasurably easier for us thanks to a small grant from the Philosophy Department and the diligent efforts of philosophy doctoral student Megan Altman, in whose debt we remain.

Our colleagues in the nascent virtual scholarly community, including many whom neither of us has ever met, have helped in numerous ways. Jim Secord at the University of Cambridge provided advice and encouragement just when it was most needed. Given the eclectic nature of many of our sources, tracking down certain references has been a challenge. Participants in the HOPOS-L listserv were helpful on more than one occasion. Though we have been based at North American institutions for the duration of this project, our research inevitably demanded that we spend a great deal of time in Argentina. Colleagues in Buenos Aires, where we spent much of the summers of 2006 and 2007 as well as the entire summer of 2009, assisted us in countless ways. We are especially grateful to Patricia Chomnalez for helping us with technical aspects of research and for her friendship. Our extended stay in 2009, during which this book reached its present form, was made possible by a grant from the Fulbright Foundation, which gave one of us (Levine) the chance to interact with colleagues and students at the Facultad de Filosofía y Letras of the Universidad de Buenos Aires. We would also like to thank the participants in Levine's graduate seminar, "Filosofía de la biología," for showing us the extent to which the issues and sources discussed in this book remain alive in Argentina today.

Our trips to Argentina have been made both productive and pleasant by the aid and comfort of many friends and family members. We are especially grateful to Susana Novoa, María Laura Novoa, Carlos J. Novoa, Natalia Laclau, Francisco Kroepfl, Inés and Manuel Novoa-Laclau, and the Bergman-Szurmuk and Chomnalez families for helping us with housing and childcare, not to mention providing the occasional welcome distraction. On three separate occasions, the hospitality of the Levine-Mendieta family in Santiago, Chile, made it possible for us to take a break. Herbert Levine exceeded family obligation by reading and commenting on several chapters and initiating us into some of the pitfalls of working with nineteenth-century German sources. At home in Tampa we relied on the support and love of our extended family. Ella Schmidt, Lucia and Mariana Stavig, Sonia Labrador, Pablo Brescia, Naomi Yavneh, Anat Pollack, and Madeline Cámara made it possible to meet dreadful deadlines and enjoy life at the same time.

Finally, we would like to thank Karen Merikangas Darling at the University of Chicago Press, who provided the suggestions that turned what was originally a somewhat different project into the present effort. It is fair to say that we would never have written this book without her support, which helped us to get past some difficult and frustrating moments. As every scholar knows, the peer review process at its best yields priceless results. Our experience showed us this process at its *very* best. Detailed reports from highly informed and diligent anonymous reviewers led to extensive corrections and revisions. We take full responsibility if errors and inaccuracies remain.

Short passages in this book have appeared previously. The story of Damiana (chap. 4) is recounted in Adriana Novoa (2009), "The Act or Process of Dying Out: The Importance of Darwinian Extinction in Argentine Culture," *Science in Context* 22, no. 2, 217–44. Our analysis of Julian Gassié's preference for Haeckel over Darwin appeared in Adriana Novoa and Alex Levine (2009), "Darwinism," in *The Blackwell Companion to Latin American Philosophy*, ed. Susana Nuccetelli et al. (Oxford: Blackwell). We are grateful to the publishers of both works for permission to use this material here.

Finally, though this project was conceived before she was, we thank Maia Levine-Novoa, who brought us back together, and to whom this volume is dedicated. Her presence in our lives has made it all worthwhile.

Note on Sources and Translations

Nearly every page of this book contains at least one quotation from a source originally published in a language other than English. Most of these originally appeared in Spanish, but significant numbers were translated from German and French and a few from Italian and Latin. We have elected, in lieu of a hundred-page appendix, to provide all such quotations in English translation only. Except where explicitly noted, all of these translations are our own.

Introduction

Man is not an improved ape; on the contrary, the apes are men that turned beastly.
— Florentino Ameghino (quoted in Podgorny 2005)

Neither apes nor humans exist, for the [Darwinian] law of progression would already have made men of the apes, and gods of men.—Santiago Estrada ([1878] 1889)

Why Darwinism? This question seems hardly worth asking, as debates concerning the scientific, philosophical, and historical significance of Darwin's contribution are as heated now, at the beginning of the twenty-first century, as they were in the middle of the nineteenth. Why Argentina? Let us be honest in admitting that, to anyone but the Argentine patriot or specialized scholar, this question seems not only worth asking but difficult to answer. This is a book on the reception of Darwinism in Argentina and on the thinkers who mediated, shaped, and exploited that reception. As such, it owes its existence to our conviction not only that both questions are worth asking but also that neither has a particularly obvious or trivial answer. Yes, the importance of Darwin's work is universally acknowledged. But in our view, the role played by analogies in the articulation of Darwinian theory, and the cultural contingency of these analogies, has never been fully appreciated. Studying how these analogies play themselves out in and are transformed by a cultural context other than that of Europe or North America does much to illuminate the ways in which the interactions between scientific analogies and culture shape the scientific enterprise. This book is thus, in part, an advertisement for the study of *peripheral science*. We believe it will persuade those interested in Darwinism of the larger importance of its reception in Argentina and, by extension, in Latin America, because by the end of the nineteenth century, this country hosted the largest and most important scientific community in the region, spearheading many scientific exchanges.

At the same time, we also believe it will persuade those interested specifically in Argentina, or more generally in Latin America, of the larger importance of Darwinism. Studies of the intellectual background to the construction of national identity, and to the emergence of social, political, and cultural institutions in nineteenth century Argentina, have typically focused on positivism. We don't want to deny the importance of this doctrine. Rather, we want to draw attention to the complementary role played by Darwinism in the articulation of positivism itself by the late nineteenth century.

This book thus has two coequal objects: the study of Darwinism, as practiced in Argentina, and the study of Argentina, as shaped by Darwinism. In this introduction, we consider each in turn before offering a brief account of the entrance made by Darwinism on the Argentine stage.

The Impact of Argentina on Darwinism: Peripheral Science and Analogy

We propose to study the vibrant scientific interaction between Europe and Latin America from the perspective of the latter. On the whole, treatments of this relationship tend to assume a strict vertical hierarchy in the flow of scientific knowledge, relegating Latin American scientists to the status of derivative thinkers. In the English-speaking world, even those who take the contributions of Latin American scientists seriously display striking limitations. These limitations, in turn, often reflect their ignorance of the importance of science in the Latin American process of nation building by the second half of the nineteenth century. Argentine science and Argentine politics marched hand in hand. The writer Paul Groussac (1848–1929) once explained why Carlos Tejedor had quit the directorship of the Buenos Aires Library by observing that in Argentina there could be no separation between intellectual and political work. He even went so far as to speculate that, had Darwin or Pasteur been Argentineans, they would have ended up as ministers or in some other high political office (Groussac 1901, 41). The scientific milieu was always quick off the mark with any theory thought to imply a solution to some political problem.

Pioneering social psychologist Carlos O. Bunge (1875–1918) was another intellectual who saw that the country's destiny would depend on science's capacity to discover the means of governing it. His theory of the unconscious and his work on the instincts both predated Freud's closely related studies. But Bunge abandoned pure psychology, dedicating himself to those disciplines he felt would allow him to better articulate the social function of science. Nor was science in Argentina the sole province of a small group of specialists. It was the subject of avid daily discussion in the popular press. In an 1882 article in *El Nacional*, Domingo Sarmiento (1811–1888), who had been the country's president from 1868 to 1874, argues that the "theory of evolution," as defined by Thomas Henry Huxley, is crucial to understanding the past. Furthermore, the most significant contributions to this theory—studies of the origins of primitive man—are taking place on the American continent (1900i, 127).[1] Two days later, in the same newspaper, Sarmiento discusses the work of Argentine naturalist Florentino Ameghino (1854–1911) and takes up his call to expand the instruction of paleontology throughout the nation. Once again following Huxley, Sarmiento insists that studies undertaken in the Americas are changing the very course of science. Supporting this work, and the new ideas emerging out of it, should thus be considered a national priority.

> In his final lecture [of the *Six Lectures to Working Men*], Mr. Huxley, the celebrated English geologist, notes the importance that paleontological investigations in America have acquired in recent years. Their significance has been such as to shake science right down to the level of genus and species. While he considers only North America, it is generally acknowledged that South America, the Pampas and Patagonia, must be of at least equal significance. And so we need workers capable of unveiling those treasures and mysteries still hidden within the earth! (Sarmiento 1900a, 128)[2]

The authors canvassed in this volume were all profoundly shaped by their perception of how living on the periphery of the scientific world limited their ability to draw scientific inferences. Both Ameghino and Bunge were acutely conscious of this peripheral status. Both lamented their lack of access—to books and other scholarly materials, to financial resources, to colleagues and correspondents, but most importantly, for our purposes, to analogies appropriate to scientific communication. Both had ample opportunity for scientific observation—Ameghino, in particular, enjoyed both geographic proximity to rich fossil beds and a younger brother skilled in their exploration—but both also felt that having grown up in Argentina had hampered their ability to formulate explanatory generalizations suitable for addressing the civilized world. Nor are they alone; a generation earlier, in 1840, Guillermo Rawson (1821–1890) cited his inability to find an appropriate analogy as the crux of his failure to articulate a theory of inheritance, while in an 1845 article he subsequently mailed to his hero Darwin, Francisco Muñiz (1795–1871), citing his great remove from "edifying society," questioned his own ability to describe his finds.[3] Bunge's and Ameghino's sense of themselves as good observers but poor crafters of analogies may help us understand Bunge's lifelong disciplinary eclecticism. It may also shed light on what American paleontologist George Gaylord Simpson, who revered Ameghino, called the latter's "inexplicable errors" (see Simpson 1984, chap. 4).

Nowhere is this consciousness of peripherality more acute than in Francisco P. Moreno (1852–1919), who at the age of 15, under the tutelage of German émigré naturalist Hermann Burmeister (see chap. 2), began amassing the vast paleontological and anthropological collections that would eventually come to serve as the scientific cornerstone of the Museo de La Plata, of which Moreno himself would become the founding director. By the 1870s, this museum had already gained substantive recognition, notably the praise of leading French anthropologist Paul Broca, who also published Moreno's account of his Rio Negro expedition in his *Revue d'Anthropologie* (Moreno 1874, 72–90). In an 1874 notice entitled "Le musée Moreno, à Buenos-Ayres," Broca reports that

> M. Moreno has recently founded in [the province of] Buenos Aires an anthropological museum, in which the collections he has assembled to date are preserved and displayed in excellent order. He has been good enough to send us four photographs depicting the main exhibition hall of this new museum. The disposition of the display cases, and the numerous objects they contain, are enough to show that this is no mere amateur collection, but one of genuine scientific value, worthy of a student of Professor Burmeister. This museum, realized in such a short time by a man brimming with such youth and ardor, is bound to grow rapidly, perhaps becoming as valuable to the study of the races of South America as [Samuel] Morton's has been, for the races of Central and North America, for the past thirty years. (1874, 375)[4]

Broca's hopes were not disappointed.[5] But the irony is that in 1880, a few years after this ringing endorsement by a leading European authority, Moreno took a leave of absence, pleading ill health, with the intention of traveling to Europe in pursuit of formal training—and the credentials and contacts that went with it. He went so far as to take classes incognito. The leading sociologist Ernesto Quesada (1854–1934), whose father had recommended the trip, reports how Broca, curious about who this man was who had so faithfully attended his recent lectures, asked to speak with him. He was naturally surprised to find the director of an important museum, and a man he himself had praised in print, enrolled in classes like any student (Barba 1977, 8).

After the term of his leave expired, Moreno requested an extension, displaying in his letter many of the insecurities common to naturalists of his generation. His consciousness of Argentina's great remove from the centers of knowledge is tempered by a zeal for the sort of prestige that might accrue to the country's credit on the strength of work like his: "The importance of the materials the Argentine Republic has to contribute to the study of the physical and social evolution of man is not unknown in Europe. But we lack the basis for comparison so essential to this kind of research. I see no choice but to seek it here [in Europe]. This is my reason for requesting the aforementioned leave" (Barba 1977, 8).

Curiously, even ironically, this same notion of peripherality has its counterpart in recent readings of canonical European thinkers. The educated twenty-first-century reader of Immanuel Kant's *Anthropology* or Georges Buffon's *Natural History* cannot help but be struck by the seeming provincialism of these towering figures of the Enlightenment (Bindman 2002). Notoriously, Kant never left Königsberg, nor was Buffon much of an explorer. Yet it was the Enlightenment, in no small measure due to Kant's contribution, whose universalism gave rise to the imperial gaze with which, as Mary Louise Pratt (1992) has argued, Europeans came to view the New World. Like Moreno, many of the intellectuals canvassed in this book, realizing the impossibility of gaining first-person access to this imperial gaze, viewed their own scientific contributions as deficient in precisely this respect. Their results could never aspire to universality. European authors would invariably do better, even at the description of South American reality.

Of course, this perception did not prevent the thinkers studied in this book from doing science, or there would have been no book to write. The character of that science was fundamentally shaped both by their self-perceptions and, relatedly, by the cultures in which they were situated. Nowhere is this more evident than in their recasting of inherited analogies or their choice of new ones. We follow Max Black, Richard Boyd, and Nancy Stepan in identifying a class of analogies that *constitute* the very sciences they help to articulate (see Black 1962; R. Boyd 1979; Stepan 1986). We believe that, as Stepan puts it, such analogies function "as the science itself—that without them the science [does] not exist" (1986, 267). In what follows, we will refer to them as "science-constitutive analogies."

To borrow a page from Michael Resnik's (1997) structuralism, we view science-constitutive analogies as *incomplete* objects: nodes within the larger patterns of scientific and extrascientific discourse, whose meanings are given by their precise position within those patterns. As the patterns shift and ramify, so does the position of a given node. Science-constitutive analogies are incomplete in the sense that the implications of such analogies are never exhausted by their initial articulation. As sciences advance, and as scientific activity expands to include people and institutions at greater geographical and cultural remove from its initial center, the significance of the analogies themselves changes.

If we are right in this, the study of peripheral science recommends itself to philosophers and historians of science as a way of addressing the contributions made to science by other sorts of cultural activity. Given that analogies arise within particular cultures, and that some analogies are science-constitutive, we would expect the familiar sciences of the nineteenth-century European academy to take on unfamiliar forms at the geographical and cultural periphery—in Argentina, for example. We would expect science itself to be transformed, in ways that need not always be limiting.

In this vein, we return for a moment to the curiously provincial, or *peripheral*, provenance of Enlightenment cosmopolitanism, Enlightenment conceptions of universality, and indeed the Enlightenment itself. Consider the universality inherent in Professor of Geography Immanuel Kant's model of the critical deployment of reason in its reflection on conditions for the possibility of knowledge. We propose that in attaining this achievement, as great as any of the Enlightenment's many intellectual contributions to posterity, Kant's own ignorance of the larger world, his very provinciality, was less a liability than an asset—or at any rate, a condition for the possibility of discovery. The man who formulated the Categorical Imperative was the son of a stern Prussian pietist; the man who distinguished, in "What Is Enlightenment?" the public and private uses of reason, was a loyal and dutiful Prussian subject and employee of the Prussian Crown (Kuehn 2001, chap. 1). In eighteenth- and nineteenth-century Europe, peculiar regional and local perspectives gave rise to philosophical and scientific schemata claiming universality and systematicity as their chief virtues. Universality, as Hegel realized, is itself historically and socially *situated*—and so, too, are various modes of particularity, of variation.

Florentino Ameghino, product of a society marginalized by an Enlightenment discourse that had been written into imperialism itself, attended not to universality, but to variation. His genius consisted in his appreciation for variability in the fossil record, his focus so acute and exclusive that, as Simpson has noted, he was wont to treat minor morphological differences between two individual specimens as grounds for naming a new species. By any standard, his taxonomy was nothing short of profligate, but for all that, his phylogeny was perhaps more Darwinian in spirit than any of its nineteenth-century rivals (see below, chap. 3, pp. 96–100). In Darwin's view, after all, heritable differences among individuals are the lynchpin of evolutionary change.

In Ameghino's hands, Darwin's Tree of Life, which German naturalist Ernst Haeckel had transformed into the "Family Tree of Man," crowned at its apex by European manhood, becomes instead a tree of death, a confusion of lopped branches. Not laboring under the confining European faith in progress, Ameghino was free to articulate an evolutionary theory that, while flawed, in many ways anticipated developments under the modern synthesis of the twentieth century.[6] It is thus no wonder that George Gaylord Simpson, one of the architects of that synthesis, so revered him. And it is also no wonder he found him perplexing (Simpson 1984). In manifest contradiction with the consensual reading of the geological record, Ameghino hypothesized that primates, including humans, originated not in Africa, but in South America. But what Simpson missed is that Ameghino's genius is inseparable from his seeming flaws; his view of primate origins cannot be taken in isolation from his understanding of the evolutionary process. This is because, for Ameghino, "Man is not an improved ape; on the contrary the apes are men that turned beastly" (Podgorny 2005, 250).[7] Science-constitutive analogies, like Kuhnian paradigms, set the parameters of scientific research programs in ways that both constrain and enable. Ameghino's contribution to scientific posterity, like Kant's, was perhaps partially enabled by the constraints of his peripheral status.

On the other side of the coin, the debates among Darwinians in Argentina, and their occasional need to "correct" Darwin when they felt they could no longer follow his lead, show culture pushing scientific analogies in very different directions. There is a profound tension between the pre-Darwinian view of the world as a place of original diversity marching toward final unity, and the Darwinian understanding of life's progression from "so simple a beginning" to its present diversity. To progress toward perfection, toward higher complexity, was to approach ultimate unity. Darwinism, to the contrary, showed that from initial unity, species diversify, following paths that cannot be assumed to lead to perfection. Final unity is simply not possible on this way of thinking.

The publication of Darwin's evolutionary work not only initiated a new way of understanding organic evolution; it also radically transformed the understanding of the relationship between humans and their environment. If nature was constantly adapting and changing, the nature of human beings was also undergoing transformation—but not necessarily tending toward perfection. Darwin, Alfred Wallace, and those who followed them did not affirm "the biological essentialism of the earlier race scientists. Darwin and his allies did just the reverse: evolution by natural selection meant that species, races, and even individuals were the creatures of contingency rather than of either design or biological essence" (Brantlinger 2003). This theory refuted central ideas of the past, including those of Alexander von Humboldt, an idol to Latin American intellectuals. In particular, it underminded the faith in a future directed by progress and harmony in nature. Following Stephen Jay Gould (2000), we mention a few of its most important consequences:

- "Nature is the scene of competition and struggle, not higher harmony. Order and good design arise by natural selection only as a side consequence of struggle, and Hobbes's 'war of all against all.' . . . The struggle, moreover, is for the reproductive success of individual organisms, not directly in the service of any higher harmony" (37).
- "Evolutionary lineages have no intrinsic direction toward higher states or greater unification. . . . The geological and climatological causes of environmental change have no inherent direction either. Evolution is opportunistic" (38).
- "Evolutionary changes are not propelled by an internal and harmonious force. Evolution is a balance between the internal characteristics of organisms and the external vector of environmental change. Both the internal and external forces have strong random components, further obviating any notion of impulse toward union and harmony" (38).

According to Gould, the changes introduced by Darwinism fundamentally altered the way in which natural history was done, so profoundly as to change the meaning of such temporal notions as "future" and "past." Human nature was defined not by altruism, but by opportunism and fitness to survive; the world became a dark place in which competition and selection challenged grounds for any optimism in the future. The notion that progress and civilization would guide the world to final unity, harmony, and perfection was important to those who, inspired by Humboldt, pursued projects of economic, social, and racial unification. Darwinian evolution, on the contrary, defended the existence of a natural law by which all organisms, including humans, underwent a constant process of diversification that would lead to the emergence of new varieties in the future. Ontologically, this meant that one could no longer assume that humans were essentially always the same, but rather that the very category of humanity was transformed in a continuous process of adaptation. In the words of David Locke, the "species is decentered. The type is not simply demoted; it disappears, and the peripheral, real, variant individuals become a major concern (for it is only on them that natural selection can operate)." More importantly, according to Locke and other authorities, Darwin was a nominalist about species, where earlier naturalists had been realists. "For him the type, if it exists at all, is just a name, a fiction of an ideal individual who carries in perfect form the various characters of the species; what is 'real' for Darwin are the variants, not the type" (Locke 1992, 177).[8]

Darwin's *Origin* also introduced completely new analogies to explain changes in nature. "For him the crucial analogy lay between the unlimited changes which could be produced by breeders of domesticated varieties on the one hand, and a similar process occurring in nature on the other" (Young 1985, 85). Influenced by Malthus, he was not as interested in progress as a force directing evolution, as in understanding civilization as the result of a population problem. We do not progress against nature with the help of civilization. Instead, we struggle against nature and survive. But the use of "anthropomorphic, voluntarist descriptions of natural selection" throughout the *Origin* introduced again the problem of analogies and their meanings. Even when seventeenth-century science had "banished purposes, intentions, and anthropomorphic expressions" from scientific explanations, this new science appeared, at moments, to rely on them (Young 1985, 93). At such moments, Darwinian ideas seemed to contextualize the rapid change of the late nineteenth century, and for this reason its analogies were used in so many different ways in the Darwinian and quasi-Darwinian literature of the time. As Robert Young has put it, one "would be hard put to find the following terms in a physics text adhering to the official paradigm of scientific explanation in terms of matter, motion, and number (or, latterly, energy and force): selection, preservation, favored, struggle (or, for that matter, the concept of life itself)" (1985, 93). Darwin himself, when clarifying the meaning of the term *struggle for existence*, said that he used it "in a large and metaphorical sense, including dependence of one being on another, and including (which is more important) not only the life of the individual, but success in leaving progeny" (C. Darwin 2006, 490).

So the problem surrounding the meaning of Darwin's theory in general, and natural selection in particular, does not originate in the writers discussed in this text. In proposing the theory of evolution by means of the mechanism of natural selection, Darwin "was not really supplying a mechanism at all. Rather, he was providing an abstract *account* at a general level of how favorable variations might be preserved. He *had* to keep his account at a certain level of abstraction since, as he confessed, he could specify neither the laws of variation nor the precise means by which variations were preserved" (Young 1985, 98; emphasis in original). Crucial for Latin American thinkers, Darwin's task "was to explain *away* the *lack* of evidence while repeatedly stressing the greater plausibility of his theory over that of special creation. Whenever he was really in trouble, he adopted the same tactic as Lyell, Chambers, and Powell had done—he appealed to the very principle which was at issue, the uniformity of nature" (Young 1985, 98). But this lack of evidence was crucial for a country like Argentina, where the discussion on heredity and population had begun well before the publication of *Origin*.

As we will show, Darwin's use of analogy will be particularly important in touching on heredity and, in the case of the human species, the biological constitution of racial categories. The Darwinian view of selection drew attention to inheritance, but with no account of the mechanisms of heredity itself; on the one hand, "the arguments of the *Origin of Species* developed the idea that variation was a fundamental attribute of all living beings. On the other hand, no mechanism either morphological or physiological was proposed to account for this fact" (Coleman 1965, 124). In Argentina this gap, which called the very future of the nation into question, was bridged by analogical thinking that responded to both scientific and patriotic needs. Race and mating became common points of reference in discussions of the reproductive demands of nation building. This led, for example, to a different use of concepts such as sexual selection. While in England the latter was practically dismissed on the grounds of its reliance on female mate choice, in Argentina, as in the rest of Spanish America, this mode of selection would become an important part of evolutionary thought. It retained this status until at least 1925, when José Vasconcelos sustained the importance of beauty in the politics of mating (Novoa 2010).

The Impact of Darwinism on Argentina

The second objective of this book is to demonstrate the importance of Darwinism in nineteenth-century Argentina and, by extension, elsewhere in Latin America. Darwinism's impact on this region's culture, from the role it played in the emergence of Spanish *Modernismo* to its role in Vasconcelos's *Raza Cósmica*, has never been fully appreciated. Though addressing this deficit would obviously go far beyond the scope of this book, we propose to make a start by considering questions beyond the obvious "How was Darwin read?" In the early twenty-first century, we find ourselves in the midst of another revolution in biological science, that of molecular genetics. A tiny percentage of the population is qualified to explain the technical details of this field, yet nearly everyone makes and understands simple references to popular versions of some of its basic concepts and consequences, such as cloning and gene therapy. Our book is about the transformative effect of another revolution in science—the articulation and dissemination of Darwinism—on society at large.

Intellectual histories of late nineteenth-century Argentina are characterized by their emphasis on the importance of positivism in structuring not only political and social life, but cultural life as well. Nor can the significance of this doctrine be denied. It is curious, however, to find relatively little mention of the Darwinian revolution in this historiographic corpus. The significance of Darwinism in the scientific and cultural life of the era becomes apparent when we examine closely the ways in which intellectuals came to terms with new, often conflicting notions of evolution and progress.

The influence of positivism is clearly in evidence among members of the so-called Generation of 1837 (those born around 1810), weaned on Comte and convinced that the application of a rational, positive method would lead to the sort of political structure necessary for civilized life. Along with positivism, the sentiments of this generation were molded by the romantic conception of civilization as the historical process of rationalization, a process destined to capture peoples everywhere in its homogenizing embrace. Along this rational trajectory toward universal assimilation, the individual would yield to the social, and the natural to the cultural.

The irruption of Darwinism with the 1859 publication of *Origin of Species* challenged not only creationism and the ecclesiastical hierarchy charged with defending it. It also challenged the belief in the very possibility of a cultural evolution independent from the biological realm.[9] While the modernizing elites of Latin America welcomed any new grounds for attacking the Church and its worldview, they were much more wary of the suggestion that progress depended on understanding and accepting the laws of nature. The newfound social importance of *natural* law in the context of the Darwinian revolution was diametrically opposed to the positivist faith in the social efficacy of *rational* law. This tension is crucial to our understanding of the situation in which intellectuals in Argentina, and elsewhere in Latin America, found themselves at the end of the nineteenth century.

Darwinism undermined the confidence in the power of a universal culture to transform the population, triumphing over nature. What's more, Argentine intellectuals, and Latin American intellectuals in general, quickly became aware that Darwin, unlike Comte or Spencer, was not interested in the construction of a synthetic theory by which to unify scientific and social thought. The author of *Origin* was not concerned with developing a philosophical system harnessed in the service of building a civilized nation. The philosophical void left by Darwin was, for a time and to a degree, filled by the work of Herbert Spencer and Ernst Haeckel, who *were* directly engaged with the application of recent biological advances to society.

The new emphasis on mating and heredity was related to new discoveries on sexual reproduction that gave rise, in turn, to what Gillian Beer (2000) has called new "plots" and new ways of imagining the future. The intellectuals canvassed in this book noticed that language and narrative played a key role in developing explanations that were based not only on physical observation, but also on crafting proper analogies. The possible plots suggested by new evolutionary ideas and the potential of those plots to serve conflicting systems simultaneously fascinated Argentine Darwinists, and, following Beer, we may speculate that our Argentine Darwinians may have understood Darwin much as she does: Darwin "was telling a new story, against the grain of the language available to tell it in. And as it was told, the story itself proved not to be single or simple. It was, rather, capable of being extended or reclaimed into a number of conflicting systems" (Beer 2000, 3).

Members of the generation that fought the wars of independence in Spanish America were convinced that the introduction of the revolutionary ideas of the Enlightenment would eventually lead to the region's homogeneity. Social, political, and racial differences would be overcome, just as free trade would unify markets. Diversity was precisely what they imagined as the chief problem in the ex-colonies, so poorly integrated under the Spanish hegemony. But this perception changed with the advent of a scientific account in which, from their unified beginnings, humanity and other species tend toward diversification. So the analogies that implied that civilization would eventually bring uniform results to diverse populations became obsolete. Instead, all species, including humans, were perceived as destined to change, to diverge over time. The analogies with which Darwin articulated his accounts of heredity and variation would become one of the most important of the many factors by which Darwinism changed Argentina.

In much the same way, the articulation of a brand of Darwinism that acknowledged the culturally contingent character of this theory led Argentine intellectuals to be bound by what, in part 2 of this book, we will call a "synthetic imperative." A theme we have broached above, and to which we will return on several occasions, is that Darwin's theory was incompatible with received philosophical systems, and Darwin himself was not interested in offering a new one in place of those he had undermined. Even in Europe, this situation left many uneasy, leading to synthetic attempts at system-building on the part of Haeckel, Spencer, and others. But in Argentina, the very project of modern nation building had been wed, since independence, to the idea of a philosophical system that tied scientific advances to social progress. It is in *this* context that synthesis becomes *imperative*.

Because nineteenth-century Argentine intellectuals labored under a synthetic imperative well before the rise of Weismannianism, neo-Darwinism, and population genetics in the Northern Hemisphere, the products of this imperative anticipated in interesting ways the twentieth-century scientific synthesis that resolved the disarray in which evolutionism had fallen. The "modern synthesis" in evolutionary theory, like the efforts of Argentine intellectuals a generation earlier, was as much a philosophical achievement as a scientific one. It too strove to remedy the metaphysical and epistemological deficiencies of Darwinism, giving rise in the process to the ontology and methodology of what Ernst Mayr (1985) has called "population thinking." In this sense, the study of Argentina introduces us to an important and vital aspect of scientific thinking, that of scientific thought at the margins, in which the constitutive elements of a theory are rearranged to suit a new environment. When such rearrangements are familiar only from within a careworn, Eurocentric, internal historiography of science, they all too easily disappear from view.

By the end of the nineteenth century, the language of Latin American political and intellectual discourse was replete with biological analogies. In this discourse, evolutionary concepts took on very particular forms. Darwinian and Haeckelian idioms were more common in science, Spencerian in philosophy. In their coexistence, we see the new science adapting to a different cultural milieu. In discussing this adaptation, this book aims to illuminate the role of the underlying analogies in evolutionary thought itself. In part 1, which consists in the first three chapters, we will review the history of the dissemination of evolutionary thought in Argentina and its reception by the intellectuals who appropriated it. In part 2, we canvass several of the most salient constituent analogies of Darwinism and their effects on and transformation within Argentine scientific and social thought. Finally, in our conclusion, we consider the nature of the synthetic evolutionary discourse that emerged in Argentina and its importance for the study of evolutionary thought in general.

PART I

CHAPTER ONE

The Roots of Evolutionary Thought in Argentina

The paleontological significance of the region that today comprises Argentina dates back to the late colonial era. In 1789 a specimen of the giant ground sloth *Megatherium* was unearthed in Luján, in the province of Buenos Aires. It made its way to Madrid, where it attracted a great deal of attention. King Charles III, in the belief that the fossil represented an extant species, ordered local authorities to find and ship him a living specimen. In Martin Rudwick's account, the bones were "assembled at the Gabinete Real (Royal Museum) by Juan-Bautista Bru (1740–1799), a conservator there." Later, in 1796, "a French official who was visiting Madrid saw the skeleton and obtained a set of Bru's unpublished plates. These were sent to the Institut [de France] in Paris, and Cuvier was asked to report on them" (Rudwick 1998, 25).

Thus, it was the young Georges Cuvier who would claim priority for his formal description of the specimen and the name of the species, which "greatly increased his personal stake in the field of fossil anatomy." Furthermore, while the megatherium itself "remained in Madrid, Cuvier's paper—published in the *Magasin encyclopédique* [of 1796] . . . made the fossil widely known" (Rudwick 1998, 26). Cuvier's conclusion, that a previously unknown terrestrial animal of such great size must be extinct, had far-reaching consequences. And though he mistakenly reported the fossil as having been unearthed in Paraguay, the sensation created by his account helped draw the eyes of the community of European naturalists in the general direction of what would soon become Argentina. A rival description by José Garriga (1796), published in Spanish shortly after Cuvier's and with more detailed engravings, gave the correct location and also attracted a great of deal international attention.[1]

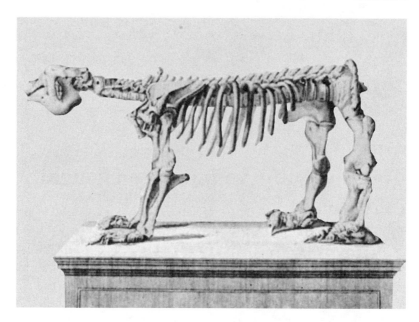

FIGURE 1. From Julian P. Boyd (1958), "The Megalonyx, the Megatherium, and Thomas Jefferson's Lapse of Memory," *Proceedings of the American Philosophical Society* 102, no. 5, 432. Original credited to José Garriga (1796), *Descripción del esqueleto de un quadrupedo muy corpulento y raro que se conserva en el Real Gabinete de Historia Natural de Madrid* (Madrid: Imprenta de la Viuda de don Joaquín Ibarra).

The discovery of the *Megatherium* put the Pampas on the scientific map. Scientists continued to be interested in the region after the separation from Spain that began in 1810 and would give rise to a formally independent Argentina with the declaration of independence of 1816. In the early days of postcoloniality in Spanish America, the study of the natural world and its evolutionary transformation was fostered by a series of European explorers. They made discoveries, ordered and classified the environment, and established lasting relationships with locals in a position to gather further specimens for European collections. In Argentina interest in the sciences also gained early support from Enlightenment-inspired leaders like Bernardino Rivadavia, who in 1812 wrote the decree creating the Museo Público of Buenos Aires, one of the first institutions of its kind in South America (Sheets-Pyenson 1988, 60–61).

From the local point of view, these expeditions were strategically important because they were associated with developing key resources. While river courses were of obvious economic significance, the geology and zoology of the region had also begun to attract significant scientific attention by the 1820s, for similar reasons. Given the expense and risk associated with mounting an expedition, beginning early in the century European naturalists were wont to encourage locals who possessed the necessary skills, resources, and inclination, to do the legwork. One such was the Buenos Aires–educated naturalist Father Dámaso Antonio Larrañaga (1771–1848), whose work was followed in France. He was born in what would later become Uruguay, where he spent most of his life and did a large part his research.[2] In an 1822 letter to Larrañaga, French explorer Louis de Freycinet (1779–1842) remarks, "The French savants wish to have the benefit of your research, and I permit myself to hope that you will direct some communication to them. Mr. Cuvier, with whom I have spoken regarding your discoveries in natural history, would be most pleased to receive an account of them. I described you to the Geographical Society as a scholar most suited to the useful advancement of the fair science to which that institution is devoted. They desire to count you among their correspondents" (Arechavaleta 1894, xxxii). In addition to Freycinet, Larrañaga would correspond with Alexander von Humboldt, Etienne Geoffroy Saint-Hilaire, and Georges Cuvier. His work promoted the interest in the natural riches of South America, particularly in areas endowed with paleontological resources.

The abundance of fossil deposits in the Pampas and Patagonia brought several important foreign naturalists to the country during the nineteenth century, Charles Darwin the most celebrated among them. In 1833 he arrived with the crew of the *Beagle*. Once in Buenos Aires, he made the acquaintance of well-connected members of the English-speaking community there, as well as meeting Juan Manuel de Rosas, the powerful political leader of Buenos Aires who would rule the entire country as dictator until 1852. At the time, as Darwin himself reported in a letter to his sister Caroline, Rosas was prosecuting "a bloody war of extermination against the Indians."[3] He was good enough to donate war-surplus horses for Darwin's exploration of the interior—and so, ironically, Rosas, who, in the received view of the late nineteenth-century Argentine establishment, was a barbarian and an enemy of science, deserves some credit for one of the most important field expeditions by the most revolutionary scientist of the age. Darwin's letter to Caroline continues by observing that "so fine an opportunity for Geology was not to be neglected." On the trail, he would "become quite a Gaucho, drink [his] Mattee & smoke [his] cigar, and then lie down & sleep as comfortably with the Heavens for a Canopy as in a feather bed.—It is such a fine healthy life, on horse back all day, eating nothing but meat, & sleeping in a bracing air, one awakes fresh as a lark."

The fact that Darwin had spent time in Argentina, done pioneering work in natural history, and made important contacts would be important in years to come. His acquaintances in the English-speaking community allowed him to stay in touch and provided a channel for communications regarding his scientific findings. One of these acquaintances, a British merchant named Edward Lumb, became a correspondent of Darwin's. According to Orione and Rocchi (1986), Darwin's ongoing contact with Lumb facilitated his correspondence with Franciso Muñiz when Darwin was searching for information about the *vaca ñata* (*ñata* oxen). In the 1840s this led, in turn, to their exchanges regarding Muñiz's saber-tooth specimen.

By the early 1830s, the results of scientific expeditions in the country were being rapidly disseminated by means of a network that encompassed not only Europe, but also Latin America. For example, in 1831, a report on an expedition by French explorer Narcisse Parchappe, an artillery officer who exiled himself to Corrientes following the Bourbon restoration and subsequently became an associate of Bonpland's and d'Orbigny's, was published in the *Revista y repertorio bimestre de la Isla de Cuba*. The Paris correspondent for the *Revista* had encountered Parchappe following his return to France in the wake of the July Revolution. He concludes that this traveler had "devoted his full attention to the geography of the regions he visited, collecting materials of great significance for improving knowledge of the Republic of Buenos-Aires, and the uses and customs of its natives. He has traced the courses of the Paraná and Uruguay, two rivers still poorly known, as well as most of the other rivers of this vast territory, as far as Patagonia. His account will be published shortly" ("Noticias" 1831, 367).

The close association of science and politics was typical of Spanish American countries in which liberal administrations, steeped in European ideas and eager to show the world their interest in progressive, modern government, facilitated and often sponsored the work of European explorers. For example, when in the 1820s and 1830s the Chilean government authorized the wide-ranging expeditions of French naturalist Claude Gay (1800–1863), one of its goals was to communicate the sense that Chile, its natural environment having been classified and therefore at least partially tamed, was progressing. Nor were the authorities wrong to think that such activity would bolster their international image. Years later, in November 1853, the "Monthly Record of Current Events" in *Harper's New Monthly Magazine* reported that for some years, "the government of Chili has devoted a good deal of attention to the progress of science. In 1833, it gave authority to Mr. Claude Gay, a French naturalist, to collect data for a political and physical history of the country: twelve volumes of which have already been published." Furthermore, "a topological survey of the country" was underway; schools were being built "for gratuitous instruction in the mechanic arts, in agriculture, painting, and music," and Chile was thus generally "acquiring high distinction among the nations for the encouragement she extends to science" ("Monthly Record" 1853). The promotion of science went hand in hand with capitalism and political liberalism; together, they were described as the solution for political and economic progress, at least in the view of the Latin Americans who promoted the supremacy of European ideas.

In an 1833 address to the French Academy of Sciences, Gay had acknowledged the importance of scientific exchange for Latin America, as well as the relationship between science and "the utility of its applications in various branches of industry." Scientific exchanges had become all the more fruitful "since the end of the last century" because of two factors, the "true philosophical spirit that now guides students even in their earliest studies" and the "numerous scientific voyages undertaken to distant countries, whether by persons attached to some government, or by private individuals" (Gay 1833, 369). Having been sponsored by the Chilean government, Gay was also in a position to comment on the relationship between science and political liberalism. Zealous though he had been when he first left France, encountering "such royal munificence on the part of a republican government raised [his] enthusiasm still further" (373).

While Latin American voyages of exploration by European naturalists earned them accolades back home, we also find that locals interested in their own natural environment were at least as attentive to the work of the scientific travelers who had set out to classify it. This attention served as Darwin's point of entry into Argentine consciousness. His account of his travels became mandatory reading, as Alexander von Humboldt's had been before him. Before the publication of the *Origin*, Darwin was thus viewed as a traveling naturalist in the tradition of Humboldt, ordering and classifying the New World. This may seem a humble role for a man destined to become the guiding light of a scientific revolution, but as this chapter will show, it was in this guise, as a participant in ongoing scientific exchanges with Argentina, a region of increasing geological and paleontological importance, that Darwin's influence continued to grow.

In terms of scientific discourse before Darwin, as Mary Pratt has explained, Humboldt "remained the single most influential interlocutor in the process of reimagining and redefinition that coincided with Spanish America's independence of Spain" (Pratt 1992, 111). Roberto González Echeverría also mentions the crucial importance that travel narratives, and mainly that of Humboldt, had in the foundation of Latin American nations. The travelers and "their writings became the purveyors of a discourse about Latin American reality that rang true and was enormously influential." The authority of this literature was "immense, not only on political developments within the very reality they described, but on the conception of that reality that individuals within it had of it and of themselves" (González Echeverría 1990, 102). For the generation who fought for independence from Spain all over Spanish America, Humboldt's views were very much the foundation of their modern nations. In 1819 Simón Bolívar delivered his famous speech to the Angostura Congress. In it he expressed the view that, above all else, the new nations needed unity.

> To save our incipient republic from . . . chaos, all our moral powers will be insufficient, unless we melt the whole people down into one mass; the composition of the government is a whole, the legislation is a whole, and national feeling is a whole. Unity, Unity, Unity, ought to be our device. The blood of our citizens is various, let us mix it to make it one; our constitution has divided authority, let us agree to unite it; our laws are the sad remains of all ancient and modern despotisms. Let the monstrous structure be demolished, let it fall, and, withdrawing from its ruins, let us erect a temple to justice, and, under the auspices of its sacred influence, let us dictate a code of Venezuelan laws. Should we wish to consult records and models of legislation, Great Britain, France, and North America, present us with admirable ones. (Bolívar 1950, 691–92; trans. in Larrazabal 1866, 371)

Bolívar's faith in unity had an important connection with Humboldt's idea on the subject. In Argentina the majority of the intellectuals born around 1810 and later known, collectively, as the Generation of 1837, relied heavily on his interpretation of the relationship between nature, science, and reason and likewise emphasized the importance of unity and harmony.[4] Humboldt's writings made a strong case for the study of these two principles in nature.

Nature considered rationally, that is to say, submitted to the process of thought, is a unity in diversity of phenomena; a harmony, blending together all created things, however dissimilar in form and attributes; one great whole animated by the breath of life. The most important result of a rational inquiry into nature is, therefore, to establish the unity and harmony of this stupendous mass of force and matter, to determine with impartial justice what is due to the discoveries of the past and to those of the present, and to analyze the individual parts of natural phenomena without succumbing beneath the weight of the whole. (1856, 24)

As is clear from this passage, for Humboldt, rational observation was ruled by the role of harmony as expressed in the union of diversity—a view that would become increasingly untenable after the publication of *Origin* in 1859, when nature became a battlefield of conflict, a place of dispute and competition.

Argentina and the Scientific Exchanges of the 1820s through 1840s

Even in the 1820s and 1830s, interest on the part of members of the more educated Argentine classes in any new idea Europe had to offer ensured that when an important scientist came to visit, his presence would not pass unnoticed. It was precisely the affirmation of the superiority of the Enlightenment ideas and secular values over those of the past that sent many young intellectuals into exile in the 1830s and 1840s. The conflict between Juan Manuel de Rosas, the conservative governor of Buenos Aires who controlled national politics, and the young generation who sought to impose a civilized culture divided Argentina until Rosas's defeat in 1852.

At the same time, and in spite of accusations against Rosas for his neglect of science, Argentina was exporting data of scientific interest. For example, the early 1830s saw the publication of two important works on the Chaco region. In 1832 Pedro de Angelis published his *Biografía del Señor General Arenales y juicio sobre la memoria histórica de su segunda campaña a la Sierra del Perú en 1821* (Biography of General Arenales and Analysis of the Historical Record of His Second Campaign to the Sierra of Peru in 1821). This account of the life and career of a prominent general of the War of Independence incorporated information on the 1821 campaign provided by his son José, including cartographic data that would be carefully read in Europe. The following year, José de Arenales (1833) himself published his *Noticias históricas y descriptivas sobre el gran país del Chaco y Río Bermejo, con observaciones relativas a un plan de navegación y la colonización que se propone* (Historical and Descriptive Notices on the Great Region of the Chaco and the Rio Bermejo, with Observations Concerning a Plan of Navigation and the Proposed Colonization). This work would become a standard reference for published accounts of expeditions in Argentine territory.

Pedro de Angelis was another foreigner who actively participated in the scientific exchanges that took place before the publication of *Origin* in 1859. He was born in Italy in 1784 and spent his early career as a journalist at home and in France, leaving for Buenos Aires in 1827 with the ostensible aim of enlisting his talents in the service of the liberal government of Bernardino Rivadavia. He later became a collaborator of Rivadavia's enemy Juan Manuel de Rosas, for whom he wrote supportive propaganda and managed historical archives. Following the account of the Chaco expedition, he was instrumental in publishing further historical documents and promoting them abroad. In 1836 he released a collection that included accounts of Luis de la Cruz's trans-Andean expedition of 1801. This book was widely distributed, attracting the notice of the prestigious *Edinburgh Review* in April 1837, especially for its accounts of the region's peoples and natural environment. "A very circumstantial account of these people [Indians] is given by La Cruz, in a memoir, inserted in the collection of Señor de Angelis.... We shall close ... repeating our persuasion, that its merits entitle it to a European as well as an American popularity; and that it is likely, by concentrating the information which relates to the interior of the American continent, in the vicinity and south of the Rio de la Plata, to direct enterprise and scientific enquiry towards it, and thus to accelerate our acquaintance with that interesting portion of the globe" ("Review of *Colección de obras y documentos*" 1837, 109).

De Angelis was also an active correspondent of important scientific and scholarly institutions elsewhere. In this position he was well aware of the significance that scientific expeditions had, and he followed closely those relevant to the country he was serving. It is not surprising, then, that he would write about the results of the Beagle expedition soon after their publication, noting departures from earlier treatments of the region. In March 1839 de Angelis cited Philip Parker King and Robert Fitz Roy in his description of the geography of Patagonia. He criticizes Thomas Falkner's (1774) *A Description of Patagonia and the adjoining parts of South America*, claiming that "the work of Mr. Fitz Roy and Mr. King, its publication just announced in England, dealing with the Patagonian coast and the Strait of Magellan," was far superior "because it is the product of many years of effective and enlightened research." He finishes the section by lamenting not having seen "the maps that the authors sent to Buenos Aires, which would have helped me to complete this notice."[5] Most importantly, as de Angelis notes, the volume announced included extracts from the *Journal* of Charles Darwin, and his description of Argentina (1839, x). This speedy recognition of the publication of a work that would become so crucial for Argentina demonstrates how the role of foreigners and their networks outside the country structured the scientific exchanges at this time.

This close contact with the European scientific community soon brought de Angelis recognition. In 1841 the Royal Geographical Society lists him as a corresponding member (lxxx). The *Proceedings of the Massachusetts Historical Society* records several items of correspondence beginning in 1841, including donations of several of his books (Historical Society of Massachusetts 1835–1855, 2:188). He was elected a corresponding member in January 1845 (303). Later that year the *Proceedings* notes with interest a letter from de Angelis containing "a continuation of the narrative of the mission of W.G. Ouseley and Baron Deffaudis" (349). Throughout de Angelis's corpus, accounts of scientific expeditions are interspersed with accounts of military campaigns, the two being inseparably linked in the discourse of both Argentine nation building and European imperialism. Throughout the 1820s, 1830s, and 1840s, the increasing interest in scientific expeditions, in both Northern and Southern Hemispheres, is thus no coincidence. The famous voyage of French naturalist Alcide d'Orbigny took place during the 1830s, as did the even more famous voyage of the H.M.S. Beagle. As the trajectories of these expeditions took them through the Pampas, they reaped a rich bounty of fossil specimens, many lying in plain sight.

In Buenos Aires, the presence of a large expatriate population, attracted by the commercial opportunities the growing port had to offer, further encouraged transatlantic scientific exchanges, in which the expatriates served as intermediaries. In the 1830s, following Rosas's rise to power, the English-speaking residents of Buenos Aires were particularly important in this regard, and none more so than Sir Woodbine Parish. Born in 1796, Parish was appointed to a consular post in Buenos Aires in 1823, and in 1825 he successfully negotiated an important commercial treaty, in recognition of which, at age 29, he was made the British Chargé d'Affaires in Argentina. As a member of the Royal Geographical Society beginning in 1824, Parish was in a position of privileged access to many prominent British naturalists. He used his appointments abroad to report and collect information about the places to which his diplomatic travels took him. In 1827, for example, he sent Atacama iron ore to the chemist Edward Turner for further research (Christianson 1837, 245). He resigned his post and returned to England in 1832 but continued to maintain many of his close Argentine contacts. Highly respected in Argentina, in England his work contributed a great deal to general knowledge of the country, and he was often consulted as an authority on matters Argentine. Darwin, for example, credits Parish in his *Journal of Researches* with informing him of a "very curious source of dispute" regarding pasture boundaries among the cattlemen of the Pampas (C. Darwin 1846, 170).

According to Parish's account, published in 1839 when he was vice president of the Geographical Society, on his arrival in Argentina in 1824, the state of knowledge in geography and related disciplines was quite poor. Requesting information from provincial governors, he was generally disappointed. General Arenales, the governor of Salta, sent him "an interesting report upon the extent and various productions of that province" and "a very fair map of it, drawn by his own son Colonel [José] Arenales," the author of *Noticias históricas y descriptivas sobre el gran país del Chaco y Rio Bermejo*, cited above. Following the ideology of his times, Parish insists on the importance that this knowledge had for future Argentines, aiding them to understand not only "the productions of their own country, but in what manner they might be rendered available in furtherance of its prosperity" (Parish 1839, x–xi).

Paleontology was not Parish's chief interest, but in the wake of the Beagle expedition, his audience's interest in Argentina was inseparable from its enthusiasm for some of the more spectacular findings. So the frontispiece of his 1839 book consists in a "representation of . . . [an] extinct monster, the Glyptodon, which has been very recently discovered at no great distance from the city of Buenos Aires, apparently in a very perfect state, and which I trust ere long will be in England. Mr. Owen, of the College of Surgeons, has been good enough to draw up for me the description of it, which I have added" (viii). While celebrating this recent acquisition, Parish confesses his regret for missing out, while in residence at Buenos Aires, on "the opportunity of making what too late I learnt would have been very acceptable additions to our zoological collections." But this complaint provides him with an opportunity to praise "Mr. Darwin, and the officers of His Majesty's ship Beagle," who "have since done much to supply these deficiencies" (xvii).

Still, Parish had not returned entirely empty-handed himself, bringing with him several significant specimens, including some *Megatherium* bones "wanting in the skeleton at Madrid, especially the bones of the tail, which singularly corroborate the anticipations of Cuvier, whose description of this remarkable monster was drawn from a representation of that specimen, the only one known to exist till mine reached Europe" (173). Parish's specimens were donated to the Royal College of Surgeons, an exhibition mounted—a notice of which was published, back in Argentina, by the *Gaceta Mercantil*—and casts distributed to museums throughout Europe, adding to the paleontological fame of the Pampas. In a report on Parish's *Megatherium*, William Clift mentions that this finding coincided with specimens found by Darwin around the same time (Clift 1835, 438).

In an address to the Geological Society given in 1837, its president, Charles Lyell, reported on the fruits of Darwin's research. Announcing that this extensive fieldwork in South America had produced important results, Lyell describes data on Chilean geology and extinct mammalia. The latter involved fossil bones found in the province of Buenos Aires and in Patagonia that had caused a great deal of excitement. Though technical descriptions of these finds had yet to be published, Lyell explains that Richard Owen had authorized him to announce "in a few words some of the most striking results which he has obtained from this examination of the specimens liberally presented by Mr. Darwin to the college of Surgeons, and of which casts will soon be made for our own and other public museums." The announcement continues with references to a skull, jaw, and teeth of the *Megatherium* and "portions of another [related] animal as large as an ox." There was also "a gigantic armadillo, as large as a Tapir." Lyell concludes that "the peculiar type of organization which is now characteristic of the South American mammalia has been developed on that continent for a long period, sufficient at least to allow of the extinction of many large species and quadrupeds" (1837, 404–5).

Darwin's successful expedition and the spectacular results it produced doubtless helped motivate Parish to write about his own experiences in Buenos Aires. Argentina had become a much more interesting place for science after the Beagle expedition, and Parish wanted to capitalize on this fact by presenting his own knowledge of the country's paleontological resources. We need to remember that the development of Darwin's work was followed with interest for a number of years. In 1835 the presentation in the Geological Society of London of some of the letters that Darwin had sent to John Stevens Henslow went extremely well. Even popular publications mentioned the fact, including *The Gentleman's Magazine*, which informed its readers that Darwin's reports included "an account of his discovery of the remains of the Megatherium" to the south of Buenos Aires (Urban 1835, 634). The level of public interest was such that, as Charles Lyell recognized a year later, "few communications have excited more interest in the Society than the letters on South America addressed by Mr. Darwin," who had devoted "four years, from 1832 to 1835 inclusive, to the investigation of the natural history and geology of South America" (Lyell 1836, 367).

In the context of this heightened interest, Parish notes that the region is likely to prove replete with finds of equal or greater significance to those already described, of which Parish mentions two. One site, near Luján, where the first *Megatherium* was found, is the property "of Señor Muniz, a medical gentleman, who was engaged in exhuming it [another *Megatherium* skeleton] with great care, and every prospect of completing the specimen." But "a still more interesting discovery is that of the apparently complete remains of another monstrous fossil animal," news of which has reached England from "Mr. Griffiths, H.M. Consul at Buenos Ayres. I trust it will not be long ere these remarkable remains are in the country, where I doubt not they will afford a rich treat to the scientific inquirer" (Parish 1839, 178b). As we have noted, the sketch sent by Griffiths served as the basis for Owen's description, included in an addendum to Parish's text, of this new species, which he called the *Glyptodon*.

The scientific curiosity of a Woodbine Parish, a diplomat who dabbled in paleontology, is doubtless unfeigned. It is likely that he also considered his scientific efforts an extension of his diplomatic duties, especially where the former offered him an opportunity to one-up the French (e.g., by filling lacunae in Cuvier's description of the *Megatherium*) or to demonstrate to the natives the British zeal for nature. Reflecting back on this period in 1852, Parish offers an interesting observation on the relative appeal, for Argentines inclined to such matters, of French and British approaches to science.

> If we can boast of Mr. Darwin's labours, a work has lately been completed in Paris under the auspices of the Government by M. Alcide d'Orbigny, a well-known zoologist, originally sent to South America to collect objects of natural history, which contains not only the results of the talented author's own observations upon those branches of science which have been his own particular study, but, like most works published under the patronage of the French Government of late years, a résumé of almost everything which has been written by

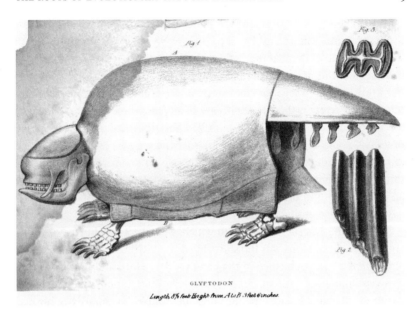

FIGURE 2. Frontispiece from Parish 1839.

others upon the countries described. Little short of an Encyclopaedia in bulk and matter, it is a splendid example of the liberality with which the Government of France is always ready to patronize and contribute to works of art and science; but it consists of eight large quarto volumes, and costs more than 50£ sterling, which is a serious impediment to its circulation and utility to the great mass of readers.

Mr. Darwin's "Journal of his Researches . . ." may be purchased for a few shillings, and a more instructive and interesting book can hardly be placed within the reach of all classes. (Parish 1852a, xxxix)

The contrast between the encyclopedic character of French scientific pursuits and the more accessible publication of Darwin certainly helps explain the rapid dissemination of Darwin's work.

The activities of European naturalists in Argentina, from titans like d'Orbigny and Darwin to relative dilettantes like Parish, had thus stimulated scientific exchange in at least four ways. First, their efforts had taught by example, an example easily enough emulated where interesting fossils lay so close to the surface. Second, they had communicated both the intrinsic value of Argentine specimens and their exchange value when offered to the right Europeans or their agents. As Parish notes in 1852, "The great interest taken by men of science in Europe in these remains was not lost upon the South Americans: I sent to Buenos Ayres the descriptions which were published of them at the time, with plates showing the parts which we possessed in this country, and those which were still wanting to complete our knowledge of these lost monsters, and I urged some of my acquaintances there to exert themselves in case of any new discoveries to endeavor to supply our deficiencies" (1852a, 218). Third, they had produced a range of texts—some of them, like Darwin's *Journal*, very accessible—in which Argentine readers could learn about their own environment as viewed through European eyes. And finally, slowly, the local press began to take an interest, both in the movements and discoveries of European naturalists and in the scientific labors of native sons.

Besides the *Gaceta Mercantil*, to which we will return shortly, one outlet for such discussions was the short-lived weekly illustrated magazine *El museo americano*, published in 1835 and 1836 by a husband-and-wife team of Swiss immigrants, César Hipólito Bacle, a lithographer, and Adrienne [Andrea] Macaire Bacle, a painter and illustrator. It primarily published translations of articles that had previously appeared in major European magazines, on topics ranging from science to fashion. But one of its original contributions was an article on the *Megatherium* by Juan María Gutiérrez (1809–1878), who would become a leading member of the Generation of 1837, occupying important posts in the nation's educational system following the fall of Rosas in 1852. His early interest in science is significant. He writes that, moved by the interest "incited by any product of one's own native soil," he wishes to offer some news "concerning an unknown or fossil animal" that "savants have given the name *Megatherium*, and of which M. Demerson says that it may justly be considered the *Hercules of the animals*" (Gutiérrez 1835–1836; repr. in B. R. Martín 2005–2006, 7–10; emphasis in original).

Gutiérrez's "news" consists, in part, in a brief but well-informed primer on paleontology in which, in addition to citing J. L. Demerson, he refers his readers to Georges Cuvier and Jean Bory de Saint-Vincent.[6] More significant, perhaps, is his reaction to the "revelation" in the *Gaceta Mercantil* of the "secret" that "*Parish* took with him, on returning to his country, the bones of a *Megatherium* far more complete, in the judgment of members of the Geological Society of London, than the famous specimen in the Madrid Gabinete de Historia Natural. . . . Our sentiments ran high indeed when we learned that an object of such great worth, to be found only in our soil, was on display in the museum of a foreign country that has never given so much as a scrap of coal to our own" (Gutiérrez 1835–1836, 9). Scientific exchange with Europe was an established practice by 1835. What Gutiérrez articulates is a pointed concern for the perceived inequality of such exchange, a veiled criticism of Rosas's lack of interest in developing a local scientific community. But expressing such sentiments was dangerous; Parish had operated with the blessing and active assistance of Rosas himself. Whether for Gutiérrez's impudence, for some comparable lapse, or for unrelated reasons, *El museo americano* was shut down by the authorities in 1836 and Bacle, its publisher, incarcerated. Though he died in 1838, his publication shows an incipient local interest in scientific news and its relationship with local politics, an association that would continue in the years to come. This local interest is even clearer with the emergence of the first important Argentine naturalist, Francisco Muñiz, a correspondent of Darwin's whose research would be published in Europe.

Francisco Muñiz, the Accidental Naturalist

Francisco Javier Muñiz was born in Monte Grande, in the province of Buenos Aires, in 1795. In 1814 he entered the Military Medical Institute (Instituto Médico Militar), then under the direction of Cosme Argerich. He completed his surgical studies in 1824, deploying to Chascomús as an army medic. In 1826 he was promoted to the rank of Army Surgeon Major, transferring in 1828 to Luján, where he served as physician to military and police units. He was a homegrown South American natural philosopher in an era when there was no institutional place for such people; no university departments or scholarly journals were available for those seeking initiation to scientific research. What Muñiz did have, however, was the unique natural wealth of the Pampas, ready to hand.

Before turning to Muñiz's paleontological research, it is worth mentioning that he was also a physician of some note and participated in scientific exchanges as both physician and naturalist. In 1833 his "A Case of Extensive Scabby Ulcerations, Cured by Vaccination," originally communicated in a letter to the Royal Jennerian Society, appeared in the *London Medical and Surgical Journal* (F. F. Muniz 1833). Muñiz's vaccination research continued throughout his career, and his correspondence with the Jennerian Society persisted through the 1840s. Charles Darwin consulted with him concerning the indigenous *ñata* cattle, reporting in the *Journal of Researches* that "Don F. Muñiz, of Luxan, has kindly collected for me all the information which he could respecting this breed" (C. Darwin 2006, 186). Darwin's contact with Muñiz was mediated by a mutual acquaintance, the wealthy English merchant Edward Lumb, mentioned above.

Muñiz's first documented work as a paleontologist dates from his early posting to Chascomús in the province of Buenos Aires. The transfer to Luján, of *Megatherium* fame, provided him with greater opportunities, which he did not neglect. During respites from the numerous military engagements in which his services were required, he hunted fossils. He financed these expeditions out of his own modest income, and they were consequently limited in scope. But by 1840 he had amassed a considerable collection. Entirely self-taught, he had no illusions as to his scientific qualifications—though he had clearly read Buffon and Cuvier, from whom he had acquired a broadly catastrophist view of natural history. His aspiration appears to have been that his collections would one day serve as the germ for a new National Museum of Natural History, an institutional focus for Argentine science. It was with this hope in mind that in 1842 he sent large crates of fossils to Governor General Juan Manuel Rosas, along with a detailed description of their contents. Rosas was by then the supreme dictator of the Argentine Federation and the only man who might have realized Muñiz's dream of a National Museum. But he was more interested in fostering foreign alliances than science and gave Muñiz's fossils to the French naval commander in the region, Rear Admiral Jean Henri Dupotet, who promptly shipped them off to Paris.

If Rosas was alert to the nuances of the scientific rivalry between England and France, he may have felt obliged to maintain parity, for the English had scored another coup the year before. Sir Woodbine Parish, we recall, had encouraged his South American associates to send whatever fossils they could lay hands on. As he would report in 1852,

> This appeal was not in vain, and after a time another interesting collection of fossil bones from the Pampas was sent to England by Don Pedro de Angelis, and purchased by the Royal College of Surgeons: amongst these remains Professor Owen detected what he did not expect, and succeeded in putting once more together in a wonderful manner the bones of a complete skeleton of another new monster hardly less extraordinary than the megatherium and glyptodon, to which he has given the name of mylodon.
>
> An entire cast of that of the megatherium is now to be seen in all its gigantic proportions in the British Museum, whilst those of the mylodon and the gyptodon stand unrivalled in the collection of anatomical treasures of the Museum of the College of Surgeons. (Parish 1852a, 218)

"Purchased by the college," the *Mylodon* had reached London, and Professor Owen's welcoming embrace, in November 1841 (Owen 1842, 3). Once prepared and placed on display, it caused a stir among a public that, it appears, was beginning to develop a taste for charismatic megafauna (see, e.g., "Notice of Fossil Sloths" 1843, 281ff; Beaumont 1843–1844, 61). Despite the excitement sparked by the *Mylodon*, given the size and significance of the Muñiz collection, the French must have felt somewhat mollified on receiving Rosas's gift.

Muñiz's hopes in Rosas having thus been dashed, he sought new ways of garnering support for his work. In the October 9, 1845, issue of the daily *Gaceta Mercantil*, he published a description of his most impressive specimen, a nearly intact fossil of a huge saber-toothed cat. It is worth noting that, though the *Gaceta* occasionally published pieces on matters of scientific interest (witness its report on Parish's *Megatherium*, discussed above), this article was hardly typical of the paper, whose pages were mostly filled with dispatches from the customs office, notices of property transactions, communiqués from foreign embassies, advertisements for everything from breweries to "dentists," and a liberal smattering of Rosist propaganda.

Muñiz's sober article is sandwiched between an inflammatory editorial condemnation of the Unitarian "savages," then locked in a power struggle with Rosas, and rather implausible claims by a local pharmacist as to the successful treatment of venereal disease.[7] It consists in several tables of measurements of the fossil specimen in question, along with comparative data for the lion, apparently derived from Cuvier. His discussion alludes to the dispute between Thomas Jefferson and Georges Buffon regarding the relative size and vigor of Old- and New-World fauna. It also cites Darwin: "I am certain that the species here reported is not among those described by the estimable *Mr. Darwin* following his fascinating exploration of the Patagonian coast and other portions of our Republic in 1832–36, and so I am the first, in the account that follows, to recommend it to the attention of savants dedicated to examining these witnesses and victims of terrible, devastating catastrophes" (Muñiz 1845; trans. in Novoa and Levine n.d.).

While Muñiz was clearly reasonably well-read, there is one embarrassment in his paper: the name given to his specimen, *Muñi-felis bonaerensis*, which was not only not a proper Linnaean binomial, but also otherwise out of sync with the general practice of scientific nomenclature. As M. Doello-Jurado has noted, this name was actually suggested by Muñiz's less biologically literate friends, and he used it only in the title of his article, "while in the description he simply calls it felis bonaerensis, this being a less objectionable denomination" (Doello-Jurado 1917, 305).

Muñiz's work gained him recognition outside his country. By the end of the 1880s, Domingo Sarmiento was called upon to prepare a posthumous edition of Muñiz's papers. In the process, one of his discoveries was the draft of Muñiz's answers to the seven questions posed by Darwin through Edward Lumb regarding the *vaca ñata* (*ñata* oxen). Of primary importance to Sarmiento was the fact that "the theory of the origin of species by means of natural selection was apparently already incubating in the mind of the audacious innovator." The significance of Muñiz in this process is indicated by the content of one of Darwin's questions, cited by Sarmiento as follows: "It is of great interest to me," says Darwin, "to learn whether the *vaca ñata* . . . resists domesticity, when exposed to civilized influence." At issue here is the important phenomenon of regression, a perennial obsession of the Latin American elites. Sarmiento goes further, however, implying that without Muñiz, and without Argentina, Darwin would never have been able to formulate his great insights. "This inquiry into the existence and later extinction of a variety of cattle raised on the ranches of Buenos Aires would be of little intellectual interest today, were it not thus linked with the celebrated theory of evolution, and the papers of Dr. Muñiz would be of lesser interest, too, except for the observations cited by Darwin in the *Voyage of the Beagle*" (Sarmiento 1885b, 257–58). Sarmiento's penchant for overstatement is well known, but it is true that Darwin cites Muñiz in his *Journal of Researches*, published in 1845. His source was apparently the same paper rediscovered, in draft form, by Sarmiento—but curiously, Sarmiento's account cites Darwin's *Journal*, rather than Muñiz's original, "because the pen of Darwin has already earned the seal of scientific approval" (259).

Taking advantage of the earlier exchange, on August 30, 1846, Muñiz mailed a copy of his *Muñi-felis* paper to Charles Darwin. Darwin's reply of February 26, 1847, is of interest in several respects. First, we may infer from Darwin's letter that Muñiz had written Darwin in part in the hope of selling his specimen abroad—not for personal gain, Doello-Jurado insists, but "to obtain some resources for the sole purpose of being thus able to prosecute his explorations in the search for fossils, as appears from copies of letters preserved in his archive" (Doello-Jurado 1917, 305). Darwin's response to this proposal (C. Darwin 1985–, 4:17–18) is either cool or simply coy; as the *Mylodon* shows, the claim that "societies only receive presents" clearly did not apply to the College of Surgeons, as Owen had freely admitted. Still, it is clear that the scientific community followed the destiny of bones from Argentina with interest. "I have lately heard from Mr. Morris that you wish to dispose of your fossil remains on some pecuniary arrangement, which I did not fully understand from your own letter to me. I have given Mr. Morris my opinion on this head, so will not here repeat it; but will only say that I conceive the only feasible plan would be to send your fossils here to some agent to dispose of them. No society will purchase anything of the kind without having them inspected, and most societies only receive presents." Through the rest of his letter, however, Darwin's tone is quite warm. He takes pains to thank Muñiz for his earlier help ("Some time since you were so kind as to send me through Mr. E. Lumb some most *curious*, and to me *most valuable*, information regarding the Niata oxen.") and to encourage future correspondence ("I should be deeply obliged by any further facts about any of the domestic animals of La Plata; on the origin of any 'breed' of poultry, pigs, dogs, cattle, etc." not to mention "the pigs, dogs, etc., which have *run wild*, and especially on the habits of these wild breeds, when their young are caught and reared . . . *any* information on all such points would be of *real service* to me.") His admiration for Muñiz's "continued zeal" seems entirely sincere. "I cannot adequately say how much I admire your continued zeal, situated as you are without means of pursuing your scientific studies and without people to sympathise with you, for the advancement of natural history; I trust that the pleasure of your pursuits affords you some reward for your exertions." And despite his tepid response to the idea of purchasing the saber-tooth—Darwin apologizes that he is "not connected with any mercantile establishment and cannot recommend agents, etc. etc."—we know that he had, in fact, attempted to help Muñiz. Two weeks earlier, in a letter to Owen, his well-connected sometime collaborator, Darwin had written,

> I have received a letter & parcel of Papers from S. F. Muniz, the gentleman who has made such wonderful collections of Fossil Bones near B. Ayres. It is to offer to the College of Surgeons various fossils, completing, as he believes, the skeletons collected by [me] . . . also an apparently nearly perfect skeleton of a new genus of carnivora, but which I have no doubt is the Machairodus. . . . But I should think it wd be highly desirable to offer to pay, if he will point out a channel, for the expences of the Boxes, the land-carriage about 20 or 30 miles, to B. Ayres, & getting them on board. If S. Muniz is encouraged, he will very probably send other things. Would it not be well to offer him copies of some of the College publications? I shall send him my Geology. . . . Wd you let me have an answer pretty soon; though I presume you will have to lay the offer before

the Museum Committee—What a grand feature a skeleton of the Machairodus would be! (C. Darwin 1985–, 4:14)[8]

One of Darwin's motives is immediately apparent. He suspects Muñiz's fossils to be of great scientific value and hopes to foster good relations with him. But lurking in the background is the perennial national rivalry between the gentleman-scientists of Britain and France. By this time Darwin knew, possibly from Muñiz himself, that a collection had been sent to Paris, and securing another such for his own country must have been an attractive prospect. He asks Owen whether he has "heard whether any collection of bones from B. Ayres has been received at Paris?? Muniz sent one by Admiral Dupotet & is *anxious* to know whether they ever arrived" (C. Darwin 1985–, 4:14). Indeed they *had* arrived, or so Darwin informs Muñiz in a postscript to his letter of February 26. Darwin also urges Owen to encourage Muñiz in other ways. "It is really very remarkable considering this man's utterly isolated position & that he must be poor, being a medical practitioner in the village of Luxan, that he keeps his zeal up: he has sent me a Spanish newspaper with a long description of the Machairodus & which I hope to get translated & if so I will send it to you. To encourage him, I shd like to get his paper in some of the Journals" (C. Darwin 1985–, 4:14). Darwin did not leave the matter there. By March 6, 1847, he had obtained the desired translation of Muñiz's paper and sent it on to Owen, to whom he writes to inquire regarding its possible publication. Such an outcome, he explains, "would greatly encourage Muniz in his search; & a S. American osteologist is a prodigy in nature" (C. Darwin 1985–, 4:23). Muñiz's peculiar combination of interests, talents, and geographical situation is, indeed, rare, as Darwin well knows. Genuine though his admiration may have been, his efforts appear to have come to naught. While two translations of Muñiz's paper on *Muñi-felis* may be found among Owen's papers, neither was ever published. Nor did Muñiz's magnificent specimen ever follow Parish's *Megatherium* and de Angelis's *Mylodon* to Britain. For over twenty years, it disappears from the historical record.

By the late 1840s, Muñiz and Darwin shared more than an interest in variation under domestication, as an 1848 paper on the ñandú shows. Noting that, among these birds, it is the males who protect and rear the young, Muñiz observes that this inversion of the "natural order" in no way undermines the species, for it is "of little importance whether the male or the female is responsible for guarding the chick. So long as the chick survives, protected from the dangers inherent to the tender, helpless condition of the young, nature will be satisfied" (Sarmiento 1885b, 142). Muñiz, too, had come to view nature as governed by a principle of survival. Such observations would lead Sarmiento to assert that Muñiz had "intuited the ideas that have begun to so exercise the modern world" (7). For our purposes it is also of considerable interest to note Muñiz's sensitivity to the importance of analogies in the practice of biological science and to the dangers of applying them carelessly. Such misapplications might serve only to disguise European ignorance of more far-flung environments. Disputing claims by Buffon, Cuvier, and others concerning the supposed similarity between ñandú, ostrich, and emu, Muñiz observes, "The analogy some naturalists have thought to establish between the African *ostrich* and the *camel*, exaggerated to the point of naming the former *Struthio camellus*, and which has been extended, in all its violence, to encompass the *ñandú*, by virtue of the resemblance of certain of its parts to those of the ostrich, strikes us, in the final analysis, as nothing more than the free play of the imagination. It is the substitution of a speculative sentiment in place of the appropriate mathematical consequence of an impartial comparative judgment" (116).

According to Sarmiento, the saber-tooth specimen was Muñiz's "favorite discovery," a find concerning which he wrote not only to Darwin, but also to "Geoffroy Saint Hilaire and the secretaries of several museums" (205). While it remained (and still remains) in Argentina, many of his other fossils may be found in Spanish, French, and Swedish museums, attesting, along with his corresponding memberships in numerous learned societies, to the extent of the scientific exchanges in which Muñiz participated. The importance of Muñiz's correspondents, Darwin chief among them, and the respects in which Muñiz had "intuited" aspects of the coming scientific revolution counted as evidence for those who would seek, with Sarmiento, to claim an Argentinean provenance for Darwinism.

Early Evolutionary Thought in Argentina

Darwinism spread quickly in Argentina. *On the Origin of Species* was published in 1859, and by 1870 it had become the subject of avid discussion in diverse forums, as we will elaborate in the coming chapters. Several factors explain the speed of its advance. Among them was the fact that science itself had gained an early foothold following Independence under the patronage of those who, inspired by the Enlightenment, saw it as the path to social and economic progress. Furthermore, paleontology was encouraged by the relative ease with which interesting specimens could be found throughout settled areas of the country. Those with a modicum of scientific training, including such physicians as Francisco Muñiz, might take to collecting and describing these materials, some of which lay scattered in plain sight. The fact that his 1845 description of what he took to be a new species was published in a mainstream newspaper, the *Gaceta Mercantil*, is evidence of the broader appeal of such research.

Also in 1845, Guillermo Rawson, destined to become a well-known physician, presented his dissertation, in which he challenged the venerable doctrine of spontaneous generation. He asserted that, while the mechanisms underlying the great chain of being were unknown, there must nonetheless be "some vital principle operating independently of the particular organs, which manifests itself differently in each organism; for two identical seeds may give rise to totally different products" (Rawson 1891b, 23). Throughout this doctoral thesis, and in later work, Rawson expounded a view of heredity that, like those of many of his contemporaries, unfolds along roughly Lamarckian lines. This Lamarckism is based, as Peter Bowler explains, "on the assumption that changes of structure produced by the activity of the adult organism can be reflected in the material of heredity and passed on to the next generation" (Bowler 2003, 243).

Rawson was "a tacit Lamarckian, not *avant-la-lettre* but *qui s'ignore*. In reading his text we cannot avoid the conclusion that he must have known Lamarck's work, at least a little" (Mañé Garzón 1990, 16; French text in original). This alleged tacit Lamarckianism was not exclusive to Rawson. Most Argentine intellectuals shared this view before the arrival of the new Darwinian evolutionism. For example, in *Recuerdos de Provincia*, published in 1850, Domingo Sarmiento defended the inheritance of acquired traits, affirming that he believed "firmly in the transmission of moral aptitude through the organs" and "in the injection of the spirit of one man into another by word and example" (Sarmiento 1913, 128–29).

Lamarck is rarely mentioned directly, but it is clear that some of his views are echoed in Rawson's and Sarmiento's work. It is difficult, though, to trace the way in which these ideas reached Argentina. At this time formal education was geared toward what were called the moral sciences. The natural sciences were secondary and known mostly from their discussion in foreign journals avidly read by those interested in European ideas. But precisely because of the absence of an educational system or standard curriculum, it is difficult to attribute a definitive influence to one or another particular source.

Rawson's contemporaries often read a singular mixture of eclectic European sources that happened to be at hand. They frequently learned of new books through journal reviews or published polemics and not the books themselves. Juan B. Alberdi, for example, may have learned via Pierre Leroux about the work of Etienne Geoffroy de St. Hilaire, whose importance should not be overlooked when considering the spread of evolutionary ideas (Ghirardi 2000, 23). St. Hilaire believed that the "external world is all-powerful in the alteration of the form of organized bodies" and that these modifications were inherited and "influenced all the rest of the organization of the animal," causing form to be "changed so as to be adapted to the new environment" (quoted in Avery 2003, 11).

But if these sources do not provide a clear picture of the genealogy of the Argentine thought they inspired, what is clear is that evolutionary thinking in Argentina was stimulated by European authors before Darwin. Sarmiento acknowledges as much when he explains that his own encounter with Darwin was prepared by his previous readings and that his thinking on the subject of evolution predated the publication of *Origin* in 1859. In 1886 he claims to have bought the sixth edition of Robert Chambers's (anonymously published) *Vestiges of Creation* in 1847 while in London. According to his narrative, the ideas presented in this book corresponded to Sarmiento's own intuitions. Finding these intuitions expressed in proper scientific language allowed him to make them his own. Sarmiento concludes by affirming that "Darwin found me well prepared. I believe in animal reason" (Sarmiento 1900k, 288).

Sarmiento's insistence on his independent intuition of the laws that guided civilization is typical of his work. In this case he was writing at a time when Darwin's brand of evolutionism had become extremely popular and wanted to show that he had followed the same path since early life. The claim may even be true. Reviewing his own intellectual development and that of other important figures of his generation, it is clear that they did not encounter Darwinism in total ignorance of other views on evolution. Sarmiento's affirmation that, from a young age, he had "viewed other animals as our companions in life, as incomplete manifestations of the common plan of creation," seems accurate enough (Sarmiento 1900k, 288). More importantly, if we analyze carefully his educational ideas in *De la educación popular*, published in 1849, we come to appreciate the importance that instruction in the natural sciences, and particularly in the concept of species, has in his plan (Sarmiento 1849, 221, 262).

We can observe the same "intuition" in the work of Juan B. Alberdi, another leading intellectual of the Generation of 1837. Alberdi, who would be responsible for the constitution of 1853, had a much more systematic education, particularly in philosophy. He began with an evolutionary perspective reminiscent of St. Hilaire and Montesquieu in its defense of environmental influence, but with the publication of *Las bases* in 1852 we see him shift toward biological reductionism. According to Bernardo Canal Feijóo, Alberdi's reading of Herbert Spencer or Darwin could not, by itself, explain his sudden emphasis on the "ineptitude of race" (Feijoo 1955, 363). Their most famous works had yet to be published, so Alberdi's shift has to be explained in terms of his own intellectual development. His readings on the subject of evolution drew on work by philosophers and naturalists, but by 1850 he had become much more interested in the work of British economists, including Malthus. So, like Darwin, he paid attention to the competitive aspect of access to resources in an approach to nature that was further removed from harmony and peace than the romanticism of many of his contemporaries. It may be no coincidence, then, that Alberdi anticipated a type of evolutionism that was more materialistic and less responsive to cultural influences.

The spread of positivism early in the 1830s was a result of the importance of the works of the Comte de Saint-Simon and Pierre Leroux, significant contributors to a school of thought that blended aspects of Idealism, the ideas of the Idéologues, and scientific materialism. We thus have evidence, in Argentina, for pre-Darwinian diffusion of an evolutionary theory compatible with an imported French vision of social transformation. Darwinism did not simply take root in bare soil. In fact, the discussion about inheritance, natural evolution, and science was quite well established among those familiar with European ideas. But because Argentine intellectuals were primarily schooled in moral philosophy, their understanding of science in general and biology in particular was disorganized. Juan B. Alberdi wrote that, thanks to their haphazard education, he and his colleagues were "ignorant [*nulos*] in the *physical and natural sciences*." According to him, the reason "was because we were only taught *moral sciences*" (Alberdi 1900c, 907; emphasis in original).

When we consider the importance that the Idéologues had in the thought of those who defended the Enlightenment in Argentina following Independence, we gain a clear sense of the radical departure the introduction of Darwinism would come to represent (Gandía 1960). Since the last years of the colonial era, the Marquis de Condorcet, for example, had been one of the main authorities for those who defended free markets. He and his followers "contemplated indefinite progress toward the complete absence of struggle among men," a prospect utterly inconsistent with Darwinism (R. M. Young 1985, 25). Condorcet believed in indefinite perfectibility through the teachings of reason and science. "Free inquiry, liberty, and justice would increasingly triumph over tyranny, superstition, and prejudice, and science provided the model for man's enlightenment. Human life would be prolonged indefinitely, and both the physical and mental constitution of man would undergo limitless improvement" (26). Interestingly, as Mary Pickering and others have noted, Condorcet was very influential in the intellectual development of Auguste Comte, showing the latter "how to combine an interest in scientific enlightenment with a concern for social reform" (Pickering 1993, 49). This would become one of the most important features of Argentine scientific thought in the future.

Rosas's ouster in 1852 made Alberdi and his peers part of a new political elite whose members had championed Enlightenment ideas from exile, in opposition to a dictator they saw as a throwback to the colonial era. Coincidentally, news of their victory reached Europe just as Sir Woodbine Parish was finalizing the second edition of his book on *Buenos Aires and the Provinces of the Rio de la Plata*. He considered this development important enough to suspend publication for the time it took him to write a postscript to his introduction, outlining political conditions as he saw them (Parish 1852a, xli–xliii). Greatly expanded, this second edition included a historical overview of the old viceroyalty, from conquest to revolution, and of the decades since independence. In preparing his revisions, Parish had drawn on assistance from English-speakers living in Argentina, whose network was evidently still in operation. In supplementing the limited information available to him through official channels, he had relied especially on a Dr. Gillies, "a Scotch physician, established at Mendoza, and Dr. Redhead, who had long been a resident at Salta—both able and willing to assist me in my research after knowledge" (xxxv).

One incentive for updating Parish's work had come in the form of a complimentary letter from "the greatest of modern geographers, Baron Humboldt, who assured me that I was [not] mistaken . . . in my hope that I had been able to add something to our general stock of geographical knowledge" (xxiv). He includes the letter, written in 1839, because it connects his work with that of Fitz Roy and Darwin. Humboldt asserts that Parish's work "and the voyage of Capt. Fitz Roy enriched by the beautiful observations of Darwin" are turning points in the history of modern geography (xix). It is also hardly surprising that the second edition includes an expanded discussion of the fossil "monsters," because the first edition, along with subsequent geological and paleontological discoveries, "had excited so much interest amongst geologists and paleontologists" (xxv).

Whereas in the first edition such matters had been confined to a brief note, now "the geological features of the Pampas of Buenos Ayres, and the remains of the extraordinary fossil monsters which have been discovered in them," appear to him "to be worthy of a separate chapter" (209). This chapter frequently cites Darwin, further cementing his position as the consummate authority on the geology of the Pampas.

In 1852, a few months after its publication in English, and a few more after the fall of Rosas, a Spanish translation of most of the second edition of Parish's *Buenos Ayres and the Provinces of the Rio de la Plata* was rushed into print. It was prepared by Justo Maeso, destined to hold numerous official posts, both at home and in the Foreign Service, in the post-Rosas Argentina. Maeso published a second volume, containing the remainder of Parish's text, over a year later and with a different publisher (Parish 1852b; 1853).

This second volume contained Parish's descriptions of the provinces of the interior. As for the delay, Maeso offers some explanation in a footnote to Parish's description of the province of Santa Fe.

> I feel obligated to inform the reader that, unable to further delay the publication of this translation, I will have to postpone until a later date my comments, notes, and additions, such as are necessary for the purpose both of clarifying and augmenting, and also rectifying Mr. Parish's text, with all the omissions and errors from which a work of this sort must inevitably suffer, having been written a long time ago, and at great remove from the subjects it describes. . . . I sincerely regret my inability to properly annotate the text at the present time, principally . . . because I lack the complete data necessary. . . . Though all fourteen [provincial] Governments acknowledge, in the most strident terms, the great importance of bringing these materials to the public, only three have so far responded to my pleas, promising to compile statistical reports. . . . The tepid response that has so delayed the completion of these important tasks is doubtless, above all, a consequence of bad habits and little practice in the great scope and utility of the data such labors yield, in every country whose government wishes it to progress at greater than a snail's pace. And yet for no country is the necessity of providing the world with reliable information about everything pertaining to it of greater urgency than for the Argentine Confederation. (Parish 1853, 10–11)

Parish's text contains errors and omissions, and his information is outdated—and yet, so Maeso implies, it remains superior to that available to the Argentine authorities themselves. "If this assertion offends the capacity and patriotism" of any of the provincial governments, he adds, they would do well to take note of the fact that Parish's "work is universally accepted in Europe as an accurate representation of the state of the Republic" (11). For the post-Rosas elite, this is unacceptable, for it is Europe that serves "as the active motor of progress in our country," and the wrong message will "divert this source of prosperity, or what's worse, the capitalist, the industrialist, or the worker who is preparing to emigrate" (12).

A translation of Parish thus serves the purpose of demonstrating, to those who now led the nation, the importance both of principles of scientific government and of communicating with the European public. The ideal response would be to make Parish's work obsolete.[9] Beginning with the enactment of the 1853 constitution, the Argentine government sponsored the work of French physician turned geographer Jean Antoine Victor Martin de Moussy (1810–1869), culminating with its commissioning of his 1860–1865 *Description géographique et statistique de la Confédération Argentine* (Geographical and Statistical Description of the Argentine Confederation) (Martin de Moussy 1860). Martin had arrived in South America in 1841. His professional success in Montevideo allowed him to become a patron of the sciences in his own right, in which capacity he gained widespread European recognition, including membership in the leading learned societies, well before the publication of this book.

The role of geography had changed with the fall of Rosas. As Maeso shows, under the new regime, the sciences were self-consciously understood as essential to improving the Argentine image in Europe, securing the country's membership in the family of "civilized" nations. Geography, in particular, was central to the modern reorganization of the country's territory. Martin acknowledges as much in his opening dedication, addressed to President Urquiza. "Europe, always insufficiently instructed, has often misjudged" Argentina. Its president had requested a complete representation of Argentina's "natural riches, and its immense resources in support of agriculture, industry, commerce, and immigration," and Martin was pleased to comply, in the service of "the strong, intelligent power that raises empires, and guides peoples down the roads of progress and civilization" (Martin de Moussy 1860, 1–2).

Interestingly, while offering the usual protestations of *personal* humility with regard to any qualifications he might have for treading such roads, Martin is unabashed in endorsing the Argentine government's choice of a Frenchman for this commission, given "the moral influence, so sympathetic and benevolent, that France exercises today throughout the various States of South America." Extolling the South American track record of his countrymen, he cites the work of "Auguste Saint-Hilaire in Brazil; Alcide d'Orbigny in Patagonia, Corrientes, and Bolivia; Bonpland in Missiones and Paraguay; Gay and Pissis in Chile; and Castelnau, Deville, and Weddell in Brazil, Bolivia, Peru, etc., etc., etc." (Martin de Moussy 1860, 3–4).

This book is meant to serve "as an exact guide in the hands of immigrants who bring to the Plata their capital and industry, a manual of immediate practical utility that teaches them, in clear and precise form, about the resources of this fertile and salubrious land, and the material progress of which it will soon be susceptible." In expressing this interest in attracting immigration, Martin acknowledges the supremacy of Europe, because, as the author affirms, "it is from Europe that the light comes, and America must necessarily draw on all the means for material, intellectual, and moral improvement accumulated there over the long experience of centuries" (Martin de Moussy 1860, 5).

For those who had defeated Rosas, repopulating the country with European immigrants was an immediate priority (see our discussion of sexual selection in chap. 5). The urgency of this goal was a consequence of scientific reasoning and European science, through whose offices alone the nation could expect to be certified as an ally of civilization and progress. With regard both to establishing and maintaining its civilized credentials and attracting European immigrants, Argentina thus relied heavily on the published opinions of foreign scientists.

Twenty years after the voyage of the *Beagle*, Darwin's accounts became doubly significant, as did those of d'Orbigny. It was their geological descriptions, more than any other reports, which helped to raise Argentina's profile in the most important scientific debates of the era. Consequently, Darwin's work on Argentina was well known to the political and intellectual elite—though by 1859 neither the *Journal of Researches* nor any of his other major works had yet been translated into French, the principle second language of Latin American intellectuals, let alone Spanish. But his most significant observations had been extensively discussed in the scientific literature, much of it in French, and some discussions had been translated—as was the case with Parish's book. Despite his own acknowledged limitations, Martin touches on a few of Darwin's insights. "Mr. Darwin's work, which justly enjoys its reputation in the scholarly world, has not been translated into French, and we infinitely regret that our incomplete knowledge of the English language prevents us from giving it the same detailed analysis we have devoted to d'Orbigny's observations and theories. We must therefore confine ourselves to a summary of the main views of this naturalist concerning the Pampean Formation" (Martin de Moussy 1860, 311). Though his authority in matters of Argentine geology would remain secure for decades to come, Darwin's reputation was about to undergo a radical transformation. The first volume of Martin's *Description* went to press shortly after the publication of *On the Origin of Species* in 1859. This interest in science and scientific expedition shows how the members of the elite that took power after the defeat of Rosas in 1852 set out to distinguish themselves from him by promoting education and the support of scientists. It was in this context that all of the most important members of the Generation of 1837 would come to express an interest in Darwin's work. Science and literature were the twin pillars of their plans to pull Argentina into the civilized world. For his generation, as the educational leader Juan Maria Gutiérrez affirmed in a public speech, literature and science were responsible for progress in human material existence and for the perfection of humanity itself (Palcos 1940, 258). In placing its stock in science, Argentina was following a pattern Nancy Leys Stepan has seen at work throughout Latin America during this period, in which "the region was involved in nationalist self-making, in which setting the boundaries between self and other and the creation of identities were increasingly carried out by and through scientific and medical discourses" (Stepan 1996, 7).

1854 saw the inauguration of a journal committed to the promulgation of science as essential to the country's development. Miguel Navarro Viola founded *El Plata cientifico y literario* with a view toward expanding the importance of civilized culture, publishing articles on the law, economics, politics, literature, and natural science. In the "Prospectus" for this journal, Navarro Viola justifies the inclusion of science by reference to its rapid growth during the first half of the century and its potential as a source of future revolutionary change. Moreover, this scientific revolution might well originate in and have a profound impact on Argentina. "We have materials at [the] disposal [of the natural sciences] that remain almost entirely unexplored: fossils, mines of all sort, and indigenous plants and animals will give rise to new mineralogy, new zoology, and new botany, and to their application ... to the healing arts" (Navarro Viola 1854, 4).

Navarro Viola clearly perceives the connections between these areas of scientific research and medicine. He also correctly anticipates the importance of Argentine materials to the coming revolution in biology, the revolution that would begin in 1859. Nor was he alone in his expectations. Argentine intellectuals of his generation closely followed scientific discussions on the origins of life and took them to heart in reflecting on the place of Argentina itself in natural history. Navarro Viola also takes approving note of the decree of May 1854 for the reconstitution of the Museo Público, a process that did not get underway until Hermann Burmeister's arrival several years later.

Navarro Viola's "Prospectus," together with other contributions to the first volume of *El Plata científico*, reveals that the way had been prepared for the spread of Darwinism well before the publication of the *Origin*. Navarro Viola and his contemporaries believed that science was "one and universal" and could only "diverge in its particular forms" (Navarro Viola 1854, 6). This belief in the unity of science would persist and, indeed, may be partly to blame for the fact some revolutionary aspects of Darwin's theory were often covered up to allow for the maintenance of a unified vision of the scientific enterprise that owed much to the French Enlightenment and Romanticism. Navarro Viola saw literature as the only hope for the nationalization of modern culture, but he saw science as the thread that bound civilized societies together.

The section of this volume devoted to natural sciences had, at thirty-six contributors, a broader base of participation than any other. Among these contributors were Aimé Bonpland and Francisco Muñiz, by then a senator. Careful attention shows that the way had been prepared for Darwinism not only by the generalized commitment to science, but by the prior fame of Darwin and his work. Muñiz, of course, had corresponded with Darwin, but by the 1850s the Englishman was generally known to the educated Argentine public for the fieldwork he had done in Patagonia and its impressive results. Articles addressed topics ranging from the treatment of mastitis and kidney stones to geology, and in one of the latter sort, dispatched from Rio de Janeiro, the names of Darwin and d'Orbigny, called "ilustres viajeros" (illustrious travelers), are cited in support of a general point about the natural history of the Earth and its transformations (de Vasconcellos and Cabral 1854, 133). The fact that such articles appear side-by-side with pieces on economic reform is also significant. The connection between natural history, biology, and economics that would become so important with the introduction of Darwinian evolutionary theory was already recognized, in Argentina as elsewhere. As John West has explained in his analysis of the business culture of the United States, the ideas of free enterprise, vigorous competition, and the *laissez-faire* approach to the role of the state in the marketplace were around long before Darwin; "they sprang from such classical economists as Adam Smith and Frederic Bastiat" (West 2005, 256). It is no coincidence that among the economic studies published by Navarro Viola in the first volume of *El plata cientifico y literario* is his own translation of an essay by Bastiat on the natural parameters of scarcity (Bastiat 1854).

A few years after Navarro Viola began publishing his journal, the first important Argentine debate on evolution, or transformism, took place. The outlet for this exchange was a popular newspaper, an indication of how scientific matters were considered of public interest, a tendency that continued in the future. In 1861 Gustavo Minelli, an exiled Italian republican and professor of universal history at the University of Buenos Aires, gave a speech denying divine creation, the original unity of the human races, and the biblical flood (on Minelli, see Montserrat 1983). This lecture—and it was lectures such as this that would earn him an excommunication by the Bishop of Buenos Aires and Pope Pius IX (Parisi 1907, 126)—caused an immediate scandal. Leading Catholic intellectual José Manuel de Estrada (1842–1894) published a reply in *La Tribuna*, a leading newspaper, which appeared again in book form the following year (J. Estrada [1862] 1899).

Estrada defended creationism against what he thought of as a Lamarckian version of evolutionism. Estrada himself was hardly a simple Biblical literalist. While defending the Biblical account of the origins of life, he also acknowledged scientific observations pointing to the variability of species. His attack on Minelli cites such secular authorities as the brothers Humboldt, Cuvier, Leibniz, and Buffon. Estrada also cites Lamarck, allowing a quotation from the *Philosophie Zoologique* to respond on behalf of Minelli with assertions in support of species transformism (J. Estrada [1862] 1899, 48).

The fact that, despite his tender years and Catholic predilections, Estrada was clearly at home with a range of scientific sources shows the extent to which members of his generation had internalized their scientific education. It is Buffon, not the book of Genesis, who serves as his primary authority in rejecting Minelli's multiple origins thesis. "The natural history of humanity," he asserts, "provides me with another powerful weapon against you, with which to prove the unity of our race." In demonstrating this claim, he boasts, "I will confine myself to scientific proofs" (37).

Estrada begins his defense of the common origins thesis by asking, "How does zoology determine what to call a species? If this name is given to any group of individuals capable of unhindered reproduction and propagation, then it cannot be denied that humanity, despite all the accidental differences among its members, constitutes a single species"(37). The closure of species as reproductive communities is, furthermore, the "barrier with which nature opposes the mixing of species" (38). Natural variations within a given species may be explained "either by physical, or by moral causes" (38). Drawing on ideas popularized by Sarmiento's generation, Estrada emphasizes the contributions of specifically environmental causes. "An African Negro, for all the degradation of his stupid savagery, is susceptible to civilization. Instruct him, exercise his intellectual faculties, and if he himself fails to display any phrenological modification, you will surely observe such advance by the second or third generation. And when his cranium has developed, his wooly hair straightened and lengthened, then I dare you to tell me that the human race is multiple, or that it had more than one original type, beautifully crafted by the skillful, omnipotent hand of an infinitely good, infinitely wise God!" (41). Estrada follows this challenge by reminding Minelli that he is not demanding that his opponent respect the Bible, only that he "respect science and history" (42). According to Montserrat, a "careful reading of this murky text allows us to infer that Darwin has not yet entered the intellectual circle in Buenos Aires" (Montserrat 2001, 2). We are not so sure about this inference. While Darwin's name is never mentioned, some of the views Estrada challenges sound much more like Darwin's than Lamarck's. Estrada charges, for example, that to ascribe to *competition* a constitutive role in human origins does injustice to our "sublime genealogy, [our] divine genesis" (J. Estrada [1862] 1899, 63). Competition, what Darwin calls the "struggle for life," is a far more important component of Darwin's evolutionary theory than it is of Lamarck's. In addition, Estrada accuses Minelli of implying that humans descend from the apes. "Instead of portraying him [man] as engaged in acts of nobility, you would show him weighed down, humbled beneath the necessities of life. You call him the King of Nature, but would name him Ape, that he might recognize his father! What a fine genealogy, that makes of Solomon a cousin to the Ass of Jerusalem!" (J. Estrada [1862] 1899, 64). If, as we believe, Estrada is here alluding to the struggle for life, the principle of natural selection, as well as the hypothesis that humans descended from the apes, then he must have some inkling of Darwinian views. Though he accuses Minelli of Lamarckianism, Estrada's own apparent belief in the inheritance of acquired traits is itself *quintessentially* Lamarckian.

In our view, Estrada's text is evidence of an important period of transition in the reception of evolutionary ideas, a period in which the constituents of evolutionary theory itself were in flux. It may be that Minelli's lecture had made broad reference to recent conceptions of evolution—including, perhaps, that of *Origin of Species*, published two years prior—all of which Estrada then attributed to Lamarck. What is certain is that this polemic, and the attendant scandal, resulted not only in Minelli's excommunication, but in the further dissemination of new scientific ideas through the medium of the popular press. The fact that a Catholic makes use of science in defense of creationism shows the great cultural significance scientific discourse had acquired by 1861.

Science was an extremely helpful tool in building the modern society, and because of this the knowledge of scientific developments became crucial for the enlightened elites. Scientific developments were also read as politically significant, relevant to the project of moving Argentina in the direction of civilization. This shift toward a secular and rationalist culture left Catholic intellectuals, such as Estrada, very concerned. Not long after his dispute with Minelli, Estrada had another famous debate with the Chilean writer Francisco Bilbao (1823–1865) over the relationship between Catholicism and democracy. Bilbao had published *La América en peligro* (America in Danger) in 1862, and the book became a success among the educated elites of Spanish America, provoking concern among Catholic authorities about their future role in a secular Argentina, a point of serious contention in the future. As in his argument with Minelli, Estrada battled fiercely against the excesses of rationalism and in favor of reconciliation between new ideas and Catholic dogma (J. Estrada 1862).

The case of these two polemics, both published in popular newspapers, illustrates another fortuitous circumstance that helped to prepare the reception of Darwinism—the perceived antagonism between the Church on the one hand, and science and rationalism on the other. This perception tended to favor those scientific ideas that appeared to secure the cause of secularization against an institution associated with the barbarism of the past. As Nancy Lays Stepan explains, Latin American intellectuals "embraced science as a form of progressive knowledge, as an alternative to the religious view of reality, and as a means of establishing a new form of cultural power. Evolution was adopted especially enthusiastically as a secular, materialist, modern view of the world" (Stepan 1996, 41).

But those who placed their faith in European evolutionary thought could not predict that, seven years after Rosas's defeat, science would undergo a revolutionary transformation that changed key assumptions, including those governing the relationship between nature and humanity. As we will explain in the next chapters, views on natural evolution will shift from a conception of nature that promoted such values as unity and harmony, to one that emphasized almost the opposite in its defense of competition and the elimination of the unfit. It is in part for this reason that the reception of Darwinism in Argentina, and its consequences, would be characterized by the ongoing tension between two opposing philosophical conceptions of evolution, hailing from two rival centers of the civilized world. The question, "Is evolution the result of a material or a spiritual principle?" would remain open and fraught.

CHAPTER TWO

The Reception of Darwinism in Argentina

Early Knowledge of Darwin's Evolutionary Views

The government's sponsorship of Martin de Moussy discussed in the previous chapter bespoke a change of official attitudes toward the role of science in public policy whose full force would not be felt until the 1860s. Legislatively mandated changes in institutions, like the rehabilitation of the Museo Público Rivadavia had founded in 1812, but which Rosas had left languishing, were slow to materialize. In 1857 Francisco Muñiz wrote to M. Trelles, secretary of the museum, to arrange for the transfer of another collection of fossils—a collection smaller than the one Rosas had sent to France, which "circumstances beyond my control had removed from my power, allowing valuable materials, intended for the Museo de la Patria, to be *sent out of the country*" (quoted in Sarmiento 1885b, 201; emphasis in original). Despite this donation—which did not include the beloved saber-tooth specimen—the museum would not find itself on a modern, scientific footing until the arrival of Hermann Burmeister, five years later.

In continental Europe, the revolutionary movements and their aftermath had given some liberal and radical intellectuals new reasons to undertake research overseas, preferably as far from Europe as possible. One of these intellectuals was the world-renowned Burmeister, a friend and protégé of Alexander von Humboldt and "long known to the world of science as one of the most learned and most active of the present generation of naturalists" ("Mammals of La Plata" 1863). From 1850 on, he spent most of his time in South America, and in 1861 he accepted the directorship of the Museo Público, having been recruited by Bartolomé Mitre and Domingo Sarmiento (Lopes and Podgorny 2000; Pyenson and Sheets-Pyenson 1999, 350–80; Sheets-Pyenson 1988, 59–68). He found the collection, which had never been professionally managed, in a somewhat chaotic state. Burmeister brought his training, expertise, and prestige to the table. It would not be too much to date the beginning of the systematic, scientific pursuit of paleontological research in Argentina to his arrival.

Shortly thereafter, he made the acquaintance of Muñiz, then in his sixties, and entered into negotiations over the *Muñi-felis* specimen. Writing in 1866, Burmeister would offer the following secondhand account of some of the exchanges related in chapter 1: "Some years before [1845], Darwin . . . had been in Buenos Aires, and entered into correspondence with Dr. Muñiz. . . . He heard of the interesting skeleton from his friend, and offered Dr. Muñiz the sum of 500 pounds sterling in order to acquire it for the British Museum. But Dr. Muñiz, so magnanimously devoted to the study of his Fatherland's fossil relics, refused to relinquish the skeleton, loudly declaring that he would transfer it only to the museum of his hometown" (Burmeister 1867, 181). Burmeister proceeds to recount having agreed with Muñiz on a sum equivalent to 1,600 Prussian Thalers, or about 240 pounds sterling. Reluctant to put what was, after all, a not inconsiderable sum on the public tab, he sought a private benefactor (D. Guillermo Wheelwright of the Central Argentine Railway), finally acquiring the specimen, for the negotiated price, on October 24, 1865. Argentina's acquisition of Burmeister and his reorganization of the museum show the interest on the part of the government in patronizing scientific activity. The implementation of this policy also coincides with an increase in specific references to Darwin's evolutionary writings in prominent journals.

Another vector for the spread of Darwinism was the continuous presence in Argentina of an active English-speaking community, within which ideas coming from England circulated quickly. As we have noted, Darwin established relationships with members of this community, such as Dr. Edward Lumb, during his stay in Argentina and continued to maintain them after his return. These connections on occasion led other Englishmen living in Argentina to consult with this leading English naturalist. One example is John Coghlan, an engineer with the railroad, who wrote Darwin in 1871 to offer his services in collecting specimens and again in 1878 to report the finding of an eight-legged horse.[1] Alfred Lumb, the son of Darwin's friend and correspondent in Buenos Aires, also continued the family tradition of facilitating scientific exchange; in 1874 *The Zoologist*

(2nd ser., 9, 4203) would report the arrival of "four rufous tinamous (Rynochotus rufencens) from the Argentine Republic, presented by Mr. Alfred O. Lumb."

But the best-known example of the influence of Darwinian thought among English-speaking residents of Argentina is that of William Henry Hudson (1841–1922), an Argentine-born son of parents from the United States. Hudson was an autodidact who first learned of the new evolutionary theory when an older brother returned from England with a copy of the *Origin*. Though he lived on the Indian frontier and had absolutely no formal education, Hudson was able to assimilate this and other scientific works, in 1867 becoming a correspondent of the Smithsonian Institution (see Arocena 2003, 37).[2] The fact that someone like Hudson, living at the outermost periphery of the scientific enterprise and with no formal training, could nonetheless gain access to the scientific establishment was another advantage of the presence, beginning in 1861, of Burmeister, who enjoyed a sterling reputation and connections with the main scientific institutions in Europe and the United States.

The connection with the Smithsonian came about through Burmeister's friend Hinton Rowan Helper, who wrote a letter to Dr. Spencer Fullerton Baird, who worked at the Smithsonian, on behalf of Hudson (Tomalin 1954, 36). Helper was the American consul in Buenos Aires and became close to the director of the Buenos Aires Museum.[3] Helper later described an encounter with Burmeister in one of his books, providing us with a good example of the kind of scientific exchanges taking place in Argentina by 1867.

> I was somewhat amused, four years ago, at my learned friend, Dr. Burmeister, the scientific German Director of the Museum of Natural History at Buenos Ayres, who came into the Consulate one day, and espying on my mantle-piece a huge tooth that had been unearthed and sent to me by another friend, Mr. Simon Ernsthal, of Cordova, in the Argentine Republic, he took it in his hands, slowly turning it over and over, and surveying it in the most thorough manner,—after which, looking at me, he remarked,
>
> "You have here, strange enough, a tooth of a species of Mastodon that has been extinct for at least fifteen thousand years." (Helper 1871, 145)

In 1866 Burmeister inaugurated the Paleontological Society of Buenos Aires, whose minutes were published in his new *Anales del Museo Público* (Annals of the Museo Público). Members of the Society were drawn from wealthy families whose interest in science was tied to their political and economic ambitions. Reports of their early meetings frequently mention Darwin in connection with discussions of species or localities he had studied. He and Richard Owen are cited, for example, in the minutes of the meeting of October 10, 1866, in an account of a recently discovered *Toxodon* specimen (Sociedad Paleontológica de Buenos Aires 1864–1869, 1:xvi). Burmeister himself delivered most such reports. The character of Burmeister's Museo Público, and of the man himself, was such as to ensure that the scientific activity of both museum and Society remained open only to those who, by dint of social class or specialized knowledge, were admitted to his select group.[4] Through Burmeister, this group had access to all the leading scientific journals, including the *Proceedings of the Linnaean Society*, in which Darwin and Wallace had published their celebrated paper of 1858. It was Burmeister who appears to have directed the course of Argentina's first serious scientific discussion of Darwin's evolutionary theory. In his latter years, and especially posthumously, Burmeister was saddled with a reputation of trenchant opposition to Darwinism in all of its forms. This view of Burmeister appears to have originated with the generation of Argentine naturalists who labored in the shadow of, and sometimes rebelled against, his authoritarian presence. In particular, Florentino Ameghino repeatedly expressed it, even while acknowledging that Burmeister's work "must be considered among the precursors of Darwinism" (1950, 79). A caricature of Burmeister as a "Biblical creationist," probably derived from Ameghino, also appears in Simpson (1984) and Montserrat (2001).[5]

This description is surely false, as is clear from the seventh German edition (1867) of his *History of Creation* and its French translation (1870). To be sure, Burmeister did resist Darwin's most provocative conclusions, as evidenced by the minutes of the July 10, 1867, meeting of the Sociedad Paleontológica: "There followed an account by Dr. Burmeister on current opinion concerning the origins of the human race, and opposing the view that it descended from the apes. By means of an analysis of the configuration of the foot, of all the parts of the human body the most particular in its construction, he demonstrated the fundamental difference between humans and apes" (*Anales del Museo Público* 1864–69, 1:xxxi). Burmeister's vast reputation in the European scientific world contributed both to the tremendous authority he wielded in Argentina and to his capacity to gain outside recognition for Argentine science (see above, p. 4, for an account of Broca's praise of Burmeister's erstwhile student Moreno) at a time when the "South American lands that provided European museums with large fossil mammals" (Lopes and Podgorny 2000, 111) had resumed funding the institutions responsible for collecting such materials, including the Museo Público. Well before his appointment as director, Burmeister had become established as "the outstanding general systematist in the developmental period in entomology" (Essig 1936, 88). His entomological handbook (Burmeister 1835) and natural history of the *Cirripedia* (Burmeister 1834) were cited extensively, the latter by Charles Darwin in his own 1854 monograph on the subject (C. Darwin 1854). His reputation did not simply evaporate with his permanent relocation to South America. In 1873 he was elected to the Entomological Society of London (*Entomologist's Monthly* 1872–73, 274).

He was also an Honorary Member of the Royal Geographic Society, which published an obituary on the occasion of his death, noting that "one hundred and sixty-four scientific papers from his pen are recorded in the Royal Society's Catalogue, and in addition he published several books of his journeys" (*Proceedings of the Royal Geographic Society* 1892, 477). Burmeister's published accounts of his travels in Brazil in 1850–1852 became well known in both Europe and the United States. In 1862 H. W. Bates wrote Darwin to comment on his criticism of Burmeister's description of the tropical forests (quoted in C. Darwin 1985–, 10:7). Burmeister's impressions of the tropics and of Brazilian slavery would become standard references for those who wrote on these topics, in both scientific and popular forums. Reputable authors would continue to refer to his descriptions of the Brazilian natural environment and of Brazilian slaves well into the twentieth century (see, e.g., P. Martin 1933).

Among the coterie of large estate holders who supported Burmeister, there was particularly strong interest in Darwin's account of natural selection. This interest may be traced in large part to the essential role played in Darwin's argument by the analogy to domestic breeding. Thomas F. Glick has studied the reception of Darwinism among ranchers in neighboring Uruguay, noting that by 1871 the members of the Rural Association, many of them foreigners, found themselves debating "whether selection was a sufficient way to upgrade the national herd" (Glick 2001, 29; for further discussion, see Mañé Garzón 1990). Sarmiento, in his study of Francisco Muñiz, describes this important function that Darwin's theory had for Argentina. By his account, selective breeding had been an essential part of the modernization of the national herds that characterized the civilization of Argentina. The elimination of the *vaca ñata* in favor of the European breeds was "achieved by taking exquisite care to cultivate and propagate the most perfect types" (Sarmiento 1900d, 192). We find similar discussions of breeding and the perfection of types in association with the replacement of the gauchos and mestizos by European immigrants.

Another fact that explains the involvement of the landowners is that, while exploring their own lands, Argentine ranchers came across intriguing fossil specimens. Those interested in studying and learning to classify them turned to Burmeister for instruction. On July 11, 1866, the first meeting of the Paleontological Society featured an analysis by Burmeister of "some bones recently found on Sr. Favier's lands" (Sociedad Paleontológica de Buenos Aires 1864–1869, 1:xi). This became a regular pattern. By the 1870s, as Irina Podgorny explains, collecting fossils was quite common outside the city of Buenos Aires. In several towns "the curiosity of teachers, priests, landowners, public employees, and doctors and pharmacists competed with those who provided fossils to European institutions specialized in natural history. It also competed with politicians from Buenos Aires interested in creating collections and with the increasing organization of the naturalist's practice in Argentina" (1997, 41).

The publication of the first issue of the *Anales* was a big event, eagerly anticipated. In an 1865 article in the widely read *Revista de Buenos Aires*, Angel J. Carranza foresaw that the journal would allow Burmeister and the museum to "establish ties with others of [their] class." Furthermore, such journals "bring honor to the nations that produce them, while at the same time setting the tone for the intellectual movements emerging in those nations" (1865, 521). Praising Burmeister for his appreciation of the importance of science in modern society, Carranza credits him with raising the level of Argentine scientific culture. Carranza further states that Burmeister's forthcoming article in the *Anales*, "Noticias preliminaries sobre los fósiles del Museo" (Preliminary Notice on the Fossil Collections of the Museum), shows Burmeister to be well equipped for the difficult work that had "opened the gates of immortality to Cuvier, Owen, d'Orbigny, and Darwin" (520).

Carranza's account shows that at this time there was no awareness of the gulf between Burmeister's and Darwin's work. In his view, both scientists were part of the same scientific current that originated in Europe. So while Darwin's name was well known to Argentine intellectuals in 1865, the revolutionary character of his evolutionary work was not completely understood. *Origin of Species* had been published in 1859, and by the mid-1860s, some sense of the stir it was causing in Europe had reached the southern shores. Carranza speaks of "a great revolution" in natural history on the horizon, but without explaining what was revolutionary in it (520).

In a November 1865 article in the *Revista de Buenos Aires*, Vicente F. López offers a fascinating comparison between the science of classical period Europeans and that of Kys-Huas and Aymara Indians. "Taking classical erudition as a point of departure," López set out to reinterpret indigenous American knowledge, promising "surprising" results. Furthermore, "the analogies we hope to extract from this procedure will justify the extravagance of our methods" (1865, 277). According to López, both classical and indigenous sources employ the same sorts of analogies in explaining the origins of life. Assuming that the classical analogies expressed secrets revealed to the initiates of the mystery cults, he asks whether "we might deduce that the American races had rested their knowledge of organic laws on a similar foundation." For an answer, he turns to Apuleius's *Metamorphosis; or The Golden Ass*.

López's reading of a central analogy (or allegory) of *The Golden Ass* is of interest to us not so much for what it reveals about indigenous scientific knowledge, but because it affords us a glimpse into López's understanding of Darwin. Apuleius's ass "attained newfound human existence after symbolically inhabiting many other animal organisms." This myth contains "profound moral and zoological truths": "Seventeen centuries after Apuleius, his myth has become the scientific basis of the great discoveries of Cuvier, Goethe, and Darwin on the relationships and links between all animal species. And it was known and taught by the Egyptians!"(281). López's text demonstrates, first of all, that he was aware both of Darwin's views and of previous evolutionary ideas. Second, his argument for the common origin of all cultures rests less on any particular event in our shared biological history than on the universality of the analogies by means of which we describe and explain that history. Third, López's analysis reveals the extent to which Darwin was catching on, not only because of the acuity of Darwin's observations, but also because of the narrative power of his hypotheses. Here evolution becomes a process of metamorphosis, known for centuries from an analogy inherent in myth. According to this analogy, we change form without losing anything in the process. Metamorphosis does not require selecting and discarding either traits or individuals.

López's interpretation of Darwin is probably indebted to the work of Jules Michelet, whose popularity in Argentina was in part responsible for the continuous spread of Lamarckianism.[6] Michelet was also interested in the analogy between evolution and metamorphosis, and his vision of nature stood in stark contrast to the seemingly bleak materialism of the *Origin*. As Edward Kaplan has argued, for Michelet, metamorphosis "bridges nature's differences. Michelet's references elsewhere to Goethe, author of *The Metamorphosis of Plants*," reinforce "Lamarck's putative spiritualism" (1977, 22). Indeed Lamarck, as Michelet understood him, sounds a lot like López. "The genius of metamorphosis had just been emancipated by botany and chemistry. It was a bold but fortunate stroke to remove Lamarck from the botanical pursuits which had occupied his life, and to impose upon him teaching about animals. This fervent genius, accustomed to miracles by the transformation of plants—full of faith in the oneness of all life—evoked the animal creation, and that great animal, the globe, from the petrification in which they had previously lain. He re-established from form to form the circulation of spirit" (cited in Kaplan 1977, 22). The failure at the time of López and many of his contemporaries to recognize that Darwinism was something very different from what Michelet and Lamarck were doing accounts, in part, for their refusal to see Darwinism as a departure from ideals of the French Enlightenment and Romanticism. Like Rawson a generation earlier, they saw evolution as an essentially benign process grounded in a spiritual principle. Michelet's "circulation of spirit" would continue to inspire an Argentine culture moved to reject a materialism that made no provision for design or progress in nature. In Argentina's reception of Darwinism, this would remain the most disturbing issue for Darwinians and anti-Darwinians alike: coming to terms with an idea of civilization whose progress toward perfection, unity, and harmony is neither intrinsic nor inevitable. Jean Baptiste Lamarck and Alexander von Humboldt would become the idols of those who opposed a strictly materialistic Darwinian account of evolution. In seeking a spiritual principle in the service of national unification, many would follow Michelet. "Society and freedom have conquered nature, history has effaced geography. In the marvelous transformation mind [or spirit] has triumphed over matter, the general over the particular, and the ideal over the real. This individual man is materialist, he willingly submits to local and private interest; human society is spiritualist, it tends ceaselessly to free itself from the miseries of local life, and to attain the lofty and abstract unity of the nation" (Kaplan 1977, 22).

In the minds of such intellectuals as López, the whole of European science and culture had to be woven together into a single narrative in which the central strand was a principle of unity, design, and perfectibility quite incompatible with the notions of struggle and variation. It would take a few more years before the truly revolutionary character of Darwinism was properly appreciated. When that happened, Darwinian evolution would consign to a world of truncated posterity and extinction all those who failed to meet a certain material condition: that of belonging to the right lineage.

Darwin makes the same sort of appearance in the *Revista de Buenos Aires* in 1868, in an article by Chilean writer and politician José V. Lastarria (1817–1888) on the geology of mountain chains. This piece follows almost word for word Jules Michelet's *The Mountain*, published earlier that same year after Michelet had read Clémence Royer's 1862 French translation of *Origin* (Lastarria 1868). Significantly, Lastarria uses Darwin in much the same way as Vicente López had three years earlier. He presents *Origin* as an updated version of theories propounded by Lamarck and Humboldt, downplaying the importance of struggle in favor of a vision of a world reaching toward unity. Lastarria follows Michelet, who had built "a case for a peaceful version of creation. His theory of knowledge implies that science and politics can reveal the same reality" (Kaplan 1977, 44). Paraphrasing Michelet, Lastarria contrasts two schools: the school of peace and the school of war. "[The school of war] is gaining ground. But the spirit of peace at all costs, which [Richard] Cobden allowed to prevail in the business of his country, appears to be what animates Lyell and Darwin. They have deemphasized the role of combat in nature, preferring that the Earth go about its business without great upheavals, transforming itself insensibly over the millions of centuries" (1868, 101). If something like this "spirit of peace" informs Lyell's geological uniformitarianism (as contrasted with Cuvier's catastrophism), the same can hardly be said of Darwin's theory of evolution, gradualistic and uniformitarian though it may be. In discussing evolution, Lastarria uses the word *metamorphosis*, where it was common at this time "to use 'metamorphosis' to express a transformist—as opposed to fixist—interpretation of species development" (Kaplan 1977, 24). Also from Michelet comes Lastarria's sense that such metamorphosis is "peaceful," and the Earth itself "beautiful" (1868, 101). Lastarria acknowledges destruction, but his sense that nature is subject to a guiding spiritual force allows his optimism to prevail.

Of course, such literary description of transformations in nature belongs to ancient tradition. In the minds of Darwin's contemporaries, the extent to which Darwin's efforts continued that tradition, as opposed to departing from it, could hardly have been immediately clear. When Vicente López, following Michelet, expresses his belief that biological evolution can be illuminated by reference to classical literary sources containing analogies similar to Darwin's, he understands Darwin as simply taking up an age-old theme. In tying the new ideas to ancient descriptions of transformation, he completely misses the fact that Darwin was not interested in explaining the great chain of being, but in the much more radical task of understanding variation in nature.

In Argentina, as in France during the same period, science and politics were inseparable. Throughout the 1860s, French Republicans made constant reference to science in their attacks on Church and Empire. This tendency became even more pronounced the following decade, in the early years of the Third Republic. The role of French Darwinism in Argentina is particularly significant because, until 1873, Clémence Royer's French translations of 1862 and 1866 were the main texts of the *Origin* accessible to those who didn't read English. This was the same translation read by Michelet. It was not until 1877, nearly eighteen years after its initial publication in English, that *Origin* appeared in Spanish.[7] *Descent of Man*, published in English in 1871, was translated into French in 1872 by Jean Jacques Moulinié, who had earlier translated Darwin's *Variation of Plants and Animals under Domestication*. When first contacted by Darwin's correspondent Carl Vogt, who at Darwin's instigation was actively searching for a replacement for Royer, Moulinié announced his fervent hope that his work would be "better than Mmle. Royer's [translation] of his book on species."[8]

Royer's translation was thoroughly impregnated with her own views and, in particular, with her appropriation of Darwin as a weapon in her attack on the Imperial establishment. In a June 1862 letter to Asa Gray, Darwin reports having recently received "a French Translation of the Origin by a Mad[elle] Royer, who must be one of the cleverest & oddest women in Europe is ardent Deist & hates Christianity, & declares that natural selection & the struggle for life will explain all morality, nature of man, politicks &c &c!!!"[9] But Darwin's enthusiasm was short-lived. By September of that same year, he expressed his frustration with the liberties Royer had taken in a letter to J. D. Hooker. "Almost everywhere in Origin, when I express great doubt, she appends a note explaining the difficulty or saying that there is none whatever!! It is really curious to know what conceited people there are in the world, (people for instance after looking at *one* cruciferous flower, explain their homologies!!!)."[10] Darwin's strong criticism of Royer's first edition led her to attempt a second, published in 1866, for which the English naturalist supplied notes. However, he was unable to rid the text of Royer's own views.[11] Her commercial success led other translators to try their hands: "J.J. Moulinié, published by Reinwald in 1873, and Edmond Barbier, also published by Reinwald, in 1876." Despite their efforts, Royer's translations remained popular, and new editions appeared in 1883 and 1918 (Clark 1981, D1028; see also Harvey 1997).

Throughout her translation, Royer had "corrected" Darwin so as to bring him in line with her political agenda. Darwin's letter points to the discontinuity between his own descriptions and analogies and those of people like Royer. Royer's alterations also imported a Lamarckian tone into discussions of natural selection, a spirit consistent with the pre-Darwinian views on evolution predominant in Argentina at the time. As rendered by Royer, Darwin's text and the analogies it contained were converted from a treatise on nature into a naturalization of the translator's politics, making her translation even more useful for some of those who sought Darwinian support for their own political agendas. According to Thomas F. Glick, Royer's interpretation conflated natural and artificial selection, while dissociating selection and struggle. Selection was presented as a "zootechnical act" in which struggle and competition had no place (1989, 38).

Readers of Royer's work "automatically encountered bits of social Darwinism." She shocked religious sensibilities "by announcing that the book challenged Christian dogma" and also provided support "for laissez-faire economics and the naturalness of inequality among races or individuals within a nation or race." She "condemned 'blind and imprudent charity' for weakening the 'human race' by its promotion of the survival of feeble individuals who, left unaided, would have perished" (Clark 1981, D1028). The influence of Royer's reading would predominate in Argentina until well into the 1870s, and Florentino Ameghino cites her several times in his 1880 monograph on the antiquity of man in the Plata region (192, 208, 209).

Another of Darwin's contemporaries who explained natural change by recourse to literary tropes, also under the influence of Royer, was the French scientist Armand de Quatrefages, whose *Metamorphosis of Man and the Lower Animals* begins with a quotation from Ovid, "Our bodies undergo transformations; we shall be to-morrow, neither what we are to-day, nor what we were yesterday" (1862; 1864, 1). Quatrefages would frequently be cited in support of the attribution of change in nature to a vital principle, rather than to strictly materialistic laws. Throughout the early decades of the reception of Darwinism in Argentina, these and other

French sources are invoked in the search for a less material, more spiritual dimension to evolution.

But if early references to Darwin's evolutionary work treated it as a continuation of accounts of nature predicated on Humboldtian romanticism, by the 1870s it had become clear what a drastic change the new evolutionism implied. In Argentina, the presence of a European authority such as Burmeister hastened this recognition. Interestingly, as we have shown, he helped to mitigate the peripheral position of Argentina through his network of scientific contacts, increasing the speed with which the country learned of debates taking place in Europe and the United States; but, on the other hand, the insulated and powerful position that Burmeister had consolidated for himself allowed him to maintain the same ideas he had defended before the publication of *Origin* without any major challenge. His work is thus a good example of what the introduction of Darwin's theory of evolution meant for established conceptions of nature, not only for Argentina, but for the European scientists who had made their careers defending them.

Darwinism and Peripheral Science: The Case of Hermann Burmeister

Karl Hermann Konrad Burmeister was born in 1807 in the Prussian Baltic port of Stralsund, until 1815 a Swedish possession (Deutsches Meeresmuseum 1993, 7–32). In 1827 he began his studies in Halle, under the direction of celebrated physician and botanist Kurt Sprengel (1766–1833). He was awarded his M.D. in 1829 with a thesis on insect taxonomy. Shortly thereafter, he gained his PhD, presenting an overview of fish anatomy. The following year he moved to Berlin, taking up a post as surgeon to the Kaiser Franz Grenadier Regiment. He apparently hoped, ultimately, to serve as a military surgeon in the tropics, perhaps in the Dutch East Indies. When this ambition was thwarted, he abandoned medical practice in favor of natural history, remaining in Berlin, in close proximity to his friend Alexander von Humboldt. He lectured in Berlin *Gymnasien*, then at the University, from 1831 to 1837, when he returned to the University of Halle-Wittenberg, first as lecturer, then professor of zoology. His charge included the University's museum, whose collections expanded greatly under his direction. 1843 saw the first publication of his great *Geschichte der Schöpfung* (History of Creation), destined to go through several editions in both German and French. In scope and spirit, the first edition of this work is in many respects a precursor to Humboldt's *Cosmos*.

The volatile year 1848 found Burmeister politically engaged, active in socialist organizations of markedly nationalist stripe.[12] In 1849 he was elected to a seat in the Prussian Herrenhaus, which he resigned in frustration the following year. That same year, 1850, with his friend Alexander von Humboldt's support, Burmeister finally fulfilled his desire to visit the tropics. He spent nearly two years in Brazil, including several weeks in the company of pioneering Danish-born paleontologist Peter Wilhelm Lund (1801–1880; see Simpson 1984). Despite a leg injury that forced him to walk with a cane for the rest of his life, and a collision at sea, he returned to Halle in 1852 with vast additions to the museum's collections in tow (see Taschenberg 1894).

Burmeister's return to Halle in 1852 took him through Paris, where he met Juan B. Alberdi, then the Argentine Confederation's Ambassador to France. Alberdi subsequently wrote President Urquiza on behalf of Burmeister, who was already planning his next voyage. This second trip, also sponsored by Humboldt and beginning in 1856, took Burmeister from Rio de Janeiro to Montevideo and Buenos Aires. In Argentina, having secured government sponsorship, he set out to explore and describe the geology of Mendoza. On his return to Buenos Aires, he purchased an agricultural estate on the banks of the Paraná, which he directed his son Heinrich Adolph to manage. This venture ultimately failed. Heinrich returned to Buenos Aires in 1859, establishing himself as a successful merchant, while his father resumed his explorations. He set out first for San Miguel de Tucumán, crossed the Andes to Chile, then sailed for Peru, returning to Europe via Panama in 1860. According to museum records, the scientific bounty of this expedition was nothing short of astounding.

To his disappointment, however, Burmeister found himself largely marginalized back in Halle (see Taschenberg 1894; Nyhart 1995, 101). Natural history had been relegated to the status of an elective in the medical curriculum, leaving his lectures virtually empty. He had gained recognition and connections in Argentina and was thus in the know when, just as his disgruntlement in Halle reached its peak, the position of director of the new Museo Público in Buenos Aires was created. French naturalist Auguste Bravard had recently declined the job (he would, in any case, die in the Mendoza earthquake the following year) and Burmeister saw his chance. After securing the sponsorship of Bartolomé Mitre and Domingo Sarmiento, he embarked for Buenos Aires in 1861. By the time he arrived, the political fortunes of these sponsors were in eclipse, and Burmeister's appointment was not confirmed for some months. Eventually, however, he set about the task of organizing the collections of the Museo and of publishing the widely distributed, scientifically rich *Anales del Museo Público* (see Andermann n.d.; Berg 1895; Biraben 1968).

It is in the first volume of the *Anales*, in 1866, that we find the first published discussion of *Muñi-felis bonaerensis*, its skeleton now rechristened *Machaerodus neogaeus*, since Muñiz's article in the *Gaceta Mercantil* over twenty years earlier (Burmeister 1864–1869). Burmeister's redescription of the specimen agrees with Muñiz in most of its measurements but corrects his nomenclature, on grounds of the priority of Cuvier's *Machaerodus*, of which Muñiz had apparently been unaware. While he was publishing the *Anales*, Burmeister's participation in European forums continued unabated.[13] The following year, Burmeister published much the same description of the specimen in the German report cited earlier, the "Bericht über ein Skelet von Machaerodus, im Staats-Museum zu Buenos Aires" (Report on a Skeleton of Machaerodus in the State Museum of Buenos Aires). While this account agrees with the Spanish text in its technical details, it is much richer in color, containing among other tidbits the account of Muñiz's interchange with Darwin.

A professional naturalist who shared Darwin's entomological interests, it appears at first glance as though Burmeister may have heard of Darwin even before the voyage of the *Beagle*. If so, he would have heard of Darwin before nearly anyone else outside of Cambridge. His 1832 *Manual of Entomology* mentions Darwin's observations of "a sand-wasp (Sphex sabulosa) which wished to carry off a large fly," shearing off the wings of its prey first (Burmeister 1832; 1836, 500). But Burmeister is here referring to Charles Darwin's well-known grandfather, Erasmus, who claimed that this example "shewed the power or reason in a wasp, as it is exercised among men" (E. Darwin 2007, 176).[14] In later work, as we have noted, Burmeister always showed profound respect for Charles Darwin the geologist, zoologist, and explorer, citing him frequently (see, e.g., 1856b, 139). His relationship with Darwin the intellectual revolutionary was much more complicated, leading to a curious layering of Burmeister's later legacy. In Germany, he had been known, and often vilified, as a radical materialist and socialist, a vocal opponent of religion, and sometime member of the "extreme left" delegation (*die äußersten Linken*) to the Herrenhaus. But as a *Naturphilosoph* of Humboldtian stripe, his approach to nature remained largely irreconcilable with Darwinism. In Argentina, this tension, coupled with his own authoritarian character, left him often at odds with younger scientists who had embraced the new theories.

True, Burmeister was no Darwinian, explicitly rejecting the hypothesis of common descent as it applied to humans in his *Der Mensch* (Man) (1868) and in later editions of the *History of Creation*. But of course this skepticism regarding the descent of humans was hardly unusual, even among figures otherwise favorably disposed to the theory of evolution by natural selection. We recall that, as noted above, one of Burmeister's chief objections, as expressed at the July 10, 1867, meeting of the Sociedad Paleontológica, concerned the anatomy of the human foot. As regards the relevance of this particular morphological peculiarty to the typological distinction between humans and apes, his convictions regarding the human foot date back at least to 1851: "We now assert that it is not the head, the hand, or the chest, but rather the leg, and most especially the foot, by which man is best distinguished, in zoological terms, from the animals. Nowhere else is man's bodily peculiarity more apparent, for besides the foot, no other body part has diverged so far from the corresponding parts of animals" (Burmeister 1851, 13). Burmeister's understanding of the taxonomic significance of the human foot had not changed in sixteen years; what had changed was the polemical context. The seventh German edition of the *History of Creation*, its final chapter updated to address the doctrine of common descent as it applied to human origins, also appeared in 1867. A French translation of this edition was published in 1870 and would be widely read in Argentina. In both texts Burmeister asserts,

> Some have been tempted to overcome this difficulty [the diversity of the human species] by recourse to the theory of species variability as advanced by Lamarck, and renewed, in more recent times, by Darwin. According to this naturalist, when a species is subjected to different external conditions, it may be gradually transformed, such that forms originally indistinguishable from one another become differentiated into numerous distinct species with their distinctive characters, over the course of geological evolution. In this way, the original specific unity of the human race has dissolved into a multiplicity of diverse types. Even the positive anatomical difference between Human and Ape feet has been described as a consequence of the modification of a primordial type, and Man himself has been viewed, in all seriousness, as a modified, perfected Ape.
>
> But we are disinclined to give our own assent to this hypothesis, though it has struck many as ingenious. As exact naturalists we must insist that problems of this sort lie outside the domain of healthy experimentation, and that it is far

better to devote oneself to what can be known scientifically, and subjected to positive scrutiny, than to become attached to conjectures that escape observation. Today, Man and Ape are both zoologically and psychologically distinct, and as we cannot abandon the principle of the invariability of species characters without overturning the whole of scientific zoology, we have every reason to believe that the differences between them are primitive, have always existed, and will persist in the future. (Burmeister 1870, 642–43)

The well-known biologist Emile Maupas executed the French translation, and the book was reviewed in some of the most important publications of the time. A review by biologist Eugène Fournier is generally representative of its French reception. While noting his differences of opinion with the German naturalist, Fournier feels obliged to do justice to "a book for which the labors of its author have earned it the respect owed to a legitimate scientific authority, conceived in an independent spirit, and without any preconceived attachment to one or another cosmogony" (1870, 136).

But Fournier takes Burmeister to task on the question of spontaneous generation, noting that, while "he cannot deny that recent experiments have rendered its existence in the present epoch extremely improbable, still he asserts that 'in the earliest ages everything was different, and so the manner of origins must also have been of different nature. Lest we take refuge in miracles and mysteries, we must concede that the appearance on the Earth of the first organized beings was the result of the free play of the generative forces of nature herself.'" In response, Fournier asserts that "at first blush it is precisely *this* hypothesis that is mysterious—that after the most delicate scientific investigations, we should still find it impossible that the production of organized beings have occurred by means of the physical and chemical forces regulating their material molecules—that would be nothing short of miraculous" (1870, 136).[15]

Fournier's review concludes by considering the curious paradox of the final chapter of Burmeister's book, "Man: the Youngest of the Earth's Creations." After citing Burmeister's insistence on the taxonomic disparity between man and the apes, Fournier remarks, "Despite this claim . . . the author here abandons logic, for he rejects the notion that all humans descend from a single couple. What's more, he insists that antediluvian humanity exhibits a different organization from postdiluvian humanity, on which point he is much closer to the Darwinists, and to Mme. Royer, than he appears to believe. And so with all due respect to the author's specialized knowledge, we have our reservations, and feel obliged to alert our readers of the consequences to which, for want of logic, this History of Creation will lead them" (Fournier 1870, 137).[16] As we have noted, by the time the edition reviewed by Fournier appeared, Burmeister had been living in Argentina for nine years and had become the leader of the anti-Darwinian front among Argentine scientists. But his arguments against common descent, as Fournier points out, were obscure and appeared to contradict some of his own views. It seems clear that he was also struggling to come to terms with how the Darwinian revolution was changing the basic paradigms that he and Alexander von Humboldt had defended for so long. By the end of the 1870s, Burmeister's view of Darwin may have evolved slightly. In 1879 he writes,

> Modern science is obliged to acknowledge original generation as the inevitable hypothesis. The observation of the remains of organized beings in the sedimentary deposits of our world reveals that primitive plants and animals were in some respects inferior to contemporary types, which must have emerged little by little, changing successively over the course of the geological epochs of their presence on this globe, until finally the arrival of man, and of the most perfect plants and animals, signals the completion of our planet's final and most sublime product.... On this basis, I am wholly convinced that the beings found in the older formations of our globe are the prototypes of contemporary beings, and in this respect, I declare myself a partisan of the hypothesis recently developed in detail, and as a natural law, by Darwin and his followers. But I must confess that their experiments have not provided me with any proof that any fundamental change in type is possible.... An insect, for example, may never transform itself into a vertebrate, because it belongs to a fundamentally opposed type. (Burmeister 1879, 11–12)

Of course, Burmeister's Darwinian "partisanship" is hardly wholehearted. To begin with, his understanding of contemporary organisms as descended from antediluvian fossils predates Darwin's *Origin* and may be found clearly expressed in his corpus at least as early as 1856. This is not a strictly Darwinian hypothesis. Furthermore, we note that despite his superficially conciliatory tone, Burmeister's objections to common descent must be understood as very deep-seated indeed. "Characters" or "types" do not, in his view, inhere only, or even primarily, in Linnaean phyla or classes, but rather in *species*. To question the possibility of "fundamental change in type" is thus to question the possibility of speciation. Indeed, like many naturalists of his generation, Burmeister was a nominalist (or conventionalist) about the higher levels of the Linnaean hierarchy, but a *realist* about species. Nowhere is this more evident than in his *Zoonomische Briefe* (Zoonomic Letters). "The only real being at hand is the final, lowermost division, called species. It alone may be seen, grasped, collected, and compiled. All of the other, higher groups are mere notions, established on the bases of one or another common feature, but whose real existence must be denied ... strictly speaking [they] have as little reality as the types they encompass; they are human creations, ideal shapes that the naturalist derives from the real forms of the species, and thus more the product of intuition than any determinate rule. This is the basis for the wavering, variable quality of the [Linnaean] system" (Burmeister 1856c, 7). Despite the 1879 passage quoted above, we find no reason to believe that Burmeister ever changed this view and plenty of evidence that he maintained it until his death in 1892.

It is important to set the record straight regarding the nature and extent of Burmeister's resistance to Darwinism, for several reasons, all related to a central theme of this book. Among other things, we are concerned with the study of peripheral science as a means of illuminating the culturally contingent aspects of scientific activity. Burmeister's intellectual peculiarities are as much cultural as they are individual. To a degree, his trouble coming to terms with Darwinism is a reflection of deep tensions between Darwin's theory and the German tradition of natural philosophy, a tradition from whose post-Darwinian currents Burmeister was at least partially cut off after his emigration.[17] In 1871 Charles Darwin quoted Burmeister's countryman Carl Vogt as asserting, two years earlier, that "no one, *at least in Europe*, any longer maintains the independent creation ... of species" (qtd. in C. Darwin 1874, 1; emphasis added). One cannot help but be reminded of two great naturalists who *had abandoned Europe for the Americas*, Louis Agassiz and Hermann Burmeister, and of their continued, increasingly insular opposition to Darwinism.

In this regard, Burmeister's persistent (if, in later years, somewhat hedged) rejection of Darwinism is of interest because of the cultural transition he himself undertook. He left Europe for Argentina as a mature thinker two years after the first publication of *Origin*, before the book was translated into French, and consequently before the community of continental naturalists had had time to assimilate its arguments. But assimilate them they would, in Germany as in France. The German champion of Darwin par excellence was Ernst Haeckel, who had much in common with

Burmeister, despite their differences. In Argentina, Haeckel's influence mounted just as Burmeister's began to decline. Given Burmeister's impact on Argentina and Haeckel's impact on the whole of Latin America, certain cultural affinities with Germany merit closer examination, along with the grounds for Burmeister's resistance to, and Haeckel's acceptance of, Darwinism. In this story, as elsewhere in this book, analogies have a big role to play.

Burmeister's published objections divide into two categories: epistemological and metaphysical. In epistemological terms, he insists that the Darwinian hypotheses be excluded from scientific discourse because they weren't subject to "positive scrutiny" (Burmeister 1870, 643).[18] We are reminded of similar scruples on the part of Darwin's countrymen William Whewell, John Herschell, and John Stewart Mill (see Hull 1989). Common descent is unverifiable; it lies outside the bounds of "healthy experimentation." His metaphysical objections amounted to a defense of species fixity as essential to the *system* of scientific zoology as he understood it: "We cannot abandon the principle of the invariability of species characters without overturning the whole of scientific zoology" (Burmeister 1870, 643).

In other words, what Burmeister clearly observed was that the new evolutionism threatened philosophical ideas at the core of what he thought was the basis of civilized science. He was unable to negotiate the resulting changes or provide a new philosophical grounding for science, because, having become himself the appointed authority on European science, he had no incentive to do so. In a country that had based its political system on the correctness and superiority of the very ideas he championed, his old view of progress and evolution remained highly congenial. Cultural contingency protected him from the evolving scientific context of the time. Had he remained in Europe, he would have been in a very different position, as were those of his colleagues he had left behind. Even had he continued maintaining the same views, he would have been pressed to address the new evolutionism more carefully, responding in a more detailed way to some of Darwin's revolutionary claims. The best illustration of Burmeister's peripheral situation may be found in a series of bitter conflicts between him and later European émigrés, scientists whom he himself helped to recruit.

In 1868, shortly before Sarmiento became president, Burmeister sent him a report in which he recommended establishing scientific institutions in the city of Cordoba, which led to the hiring of several foreign professors.

The Academia de Ciencias (Academy of Science) was founded in 1869, and after the newly recruited academics began to arrive, Burmeister was named its first president in 1873. In this capacity, he drafted its original by-laws in 1874. During this period, Burmeister brought to the country important figures in chemistry (Max Siewert from Halle University), botany (Pablo Lorentz from the University of Munich), geology (Alfred Stelzner from the Bergakademie of Freiberg), and zoology (Hendrik Weyenbergh, Utrecht and Göttingen Universities) (see Leanza 1992, 392).

But just as Burmeister's hold on the scientific establishment was reaching its peak, extending into new regions and institutions, Darwinian ideas were spreading, both in Europe and Argentina. The new professors of zoology and botany were hired from institutions with which Burmeister was eminently familiar, but they had been forced to participate in the latest evolutionary debates and had found ways to reconcile received philosophical premises with the new reality of post-Darwinian science. In the 1860s and early 1870s, while Burmeister had been shielded by his peripheral position, they had found themselves in the thick of the redefinition of science that was taking place in Europe. But to their surprise and amazement, the very colleague who had recommended their appointment became an implacable enemy, unwilling to forgive their betrayal of strict pre-Darwinian tradition. Because Darwin's influence was strong in zoology, this field was the target of Burmeister's particular scrutiny.

In an 1875 issue of the *Periódico Zoológico* (Zoological Journal), the president of the Zoological Society, Hendrik Weyenbergh, alludes to a confrontation "with a man who ought to have been [the society's] support and strength, but who instead has sought by all means to destroy it" (1875b, 3).[19] In this same issue, Weyenbergh also introduces the first article written by a native-born Argentine, the young Eduardo Holmberg, and uses the occasion to attack Burmeister, revealing an obvious generational and cultural conflict with the younger scientists who had been weaned on the debates sparked by the Darwinian revolution. "I have never agreed with Dr. Burmeister, who believes that the sons of the Argentine Republic are incapable of serious work in the exact sciences. Dr. Burmeister has often told me that all Argentines are inept boys, from whom no work in the natural sciences can be expected. He speaks in the same terms of disdain of all the eminent men of the Republic" (Weyenbergh 1875a, 277–78).

Clearly the new generation of promising native scientists was interested in Darwin, a proclivity that would hardly raise their stock in the eyes of a Burmeister. Weyenbergh and the journal that he directed encouraged this interest, and he directly cites the new theory in this same issue. In an article on recent zoological work on South America, he refers to the "decisive importance of Darwin's theory" in a discussion of migration (1875c, 307). Burmeister's role in this debate and his battles with other scientists cemented his growing reputation as interested only in his European contacts, an aristocrat dismissive of local talent. His limited engagement with Darwinism also contributed to this image of a man out of touch with Argentine reality (for more information, see Acosta 2006).

Ironically, in the world he had left behind, Burmeister's legacy was being read rather differently. In Europe his reputation as a materialist and opponent of religion was so strong that Father Ceferino González, in his *Historia de la Filosofía* (History of Philosophy), published in Spain in 1886, branded Burmeister a "radical Darwinian" aligned with the Darwinian evolutionist Ernst Haeckel (288). He also classified him as a member of the "Hegelian left" with, among others, Émile Littré, Thomas Henry Huxley, and Paolo Mantegazza (229). We can only imagine the horror that Burmeister would have felt at being placed in such company, but this example shows the degree of confusion to which evolutionary ideas had given rise. In fact there *were* many common elements between Burmeister and Haeckel, because Haeckel's evolutionary phylogeny was strongly influenced by Burmeister's approach to systematic zoology. More importantly, an idealist conception of type was important to both of them in their understanding of racial differences.

But by the 1860s Haeckel had already been recognized as having outdone Burmeister even in his own favorite field of morphological systematics (Jaeger 1871, 282). What's more, this accomplishment was widely credited to Haeckel's willingness, *contra* Burmeister, to take Darwinian phylogeny into account in taxonomy. For example, in an 1869 article on recent advances in the study of rotifers, Samuel Bartsch notes that the taxonomic place of the rotatoria had long been a point of contention among zoologists, with "Ehrenberg . . . placing them among the infusoria, while Burmeister and Leydig did everything they could to find them a place among the crustaceans." Now, with Haeckel's help, "they have found their true place, thanks to the theory of common descent" (Bartsch 1869, 325).

But despite the brief refutation of Darwinism in the seventh German edition (1867) and first French edition (1870) of the *History of Creation*, neither volume contains a single mention of Haeckel, an omission too remarkable to put down to mere oversight (though a cursory reference to Haeckel may be found in Burmeister 1879, 24). For his part, Haeckel frequently refers to Burmeister, and often—though not always—in complimentary terms. One such reference occurs in Haeckel's own great contribution to morphology, his *Generelle Morphologie der Organismen* (General Morphology of Organisms; 1866). In short order, Haeckel waves aside both Burmeister's metaphysical objection (his insistence on species fixity as constitutive of systematic zoology) and his epistemological objection. Reflecting on the pre-Darwinian attitude of naturalists toward the species category, Haeckel observes, "Until recently, the majority of naturalists, even when prepared to admit a certain degree of arbitrariness [in classification] nonetheless treated the species-concept as an exception. The species-category alone was supposed to constitute an absolutely determinate, real, clearly defined sum of forms, grounded in nature herself" (1866, 378). As an exemplary advocate of this now outdated view, he cites Burmeister.

> This conception both of the [taxonomical] system and of its various categories, which more or less dominates the understanding of most zoologists and botanists, and is applied throughout systematics, is expressed most clearly by Burmeister, a systematist who distinguishes himself from many others by virtue of his clarity and perspicacity, in his *Zoonomische Briefe*. Like Linnaeus, he compares the customary categories of the schema we have just described with the hierarchy of groupings in an army. The phyla, classes, orders, families and genera of the animal and plant kingdoms, like the divisions, regiments, battalions, companies, platoons, and squads of an army, are mere notions, ideal abstractions, meaningful only by virtue of the plurality of real bodies, the individuals that underlie them. In an army, these individuals are particular soldiers. According to Burmeister, in the organic system, they are the species. He says, 'The only real being at hand is the final, lowermost division, called species. . . .' Burmeister defends this view in detail . . . , and his exposition is especially noteworthy because it so clearly reveals the prejudice with regard to the species-concept under which this superlative systematist labors, one who plies the systematic trade with more sense and understanding than most others. (1866, 378–79)

But Burmeister's problem isn't just that he continues to adhere to the doctrine of species fixity, made obsolete by Darwin. In doing so, Haeckel accuses, he fails to follow through on *the logical consequences of an analogy* he himself has accepted: "If Burmeister, in his treatment of the very revealing comparison between the systematic categories and the hierarchical groupings of an army, locates the real individual, corresponding to the individual soldier, at the level of the species, then he has taken a great step backwards even from Linnaeus, in whose schema *miles* [soldier] is aligned with *individuum*" (379; emphasis added). Analogies carry scientific weight, and Burmeister, Haeckel charges, has paid insufficient care to entailments of his own analogies. Seen in this light, Burmeister's insistence on positive evidence, like William Whewell's insistence on the canons of inductive reasoning, is a perfectly intelligible response to a scientific theory based—more, perhaps, than any other—on *arguments from analogy*.

Because, as we argued in our introduction, analogies and the scientific arguments based on them must be understood as contingent on the cultural contexts in which they are articulated, Burmeister's situation is a good case in point. As the designated champion of a certain brand of the Enlightenment in Argentina, he saw its defense as essential to the preservation of science itself. By contrast, homegrown scientists were free to blend old and new ideas without the concerns that affected Burmeister. Regardless of the degree of his intellectual dogmatism, within two decades of his arrival the scientific culture of the country had been transformed. Writing in 1888, Emilio Daireaux offers the following observations on Burmeister and his disciples:

> A school . . . has arisen in pursuit of yet another class of discovery. Its members seek to unearth the country's historical secrets, and include anthropologists, like Don Francisco Moreno; ethnographers, like Estanislao Zeballos; and explorers, like Lista and Zeballos. These men have followed in the footsteps of those illustrious savants who have explored this country and its neighbors since the beginning of the century, whose names include Humboldt, Bonpland, D'Orbigny, Darwin, and Bravard. . . . The distinguished members of this new school, to whose excellence their own work testifies, all owe a great deal to this eminent scholar [Burmeister], who has shown by his own example the path by which we might follow our illustrious forebears, and who with his own books, in which he relates his daring explorations, has contributed so much to the understanding of our national origins, and of prehistoric races. (1888, 431)

Burmeister had brought German systematicity to Argentina—though, as all of the Argentine disciples Daireaux cites were well aware, the younger Haeckel had outdone him on this very score. Haeckel had integrated Darwin's insights into a philosophical system eminently congenial with the predilections of midcentury Argentine intellectuals. He forged a philosophical continuity between the world before and after Darwin. We will return to consider the ultimate fate of this system in later chapters, and particularly to the ways in which its ability to connect politics and science made him a key figure in the process of making Darwinian ideas so integral to Argentine thought.

The Understanding of Darwinian Evolutionism

As is evident from the debates discussed in the previous section, by the 1870s it was clear that Darwinian ideas were no mere continuation of the old tradition of natural philosophy or French Enlightenment ideas, but a radical departure from them. Recognition of this fact left the whole conception of science as an expression of a universal culture inherent in civilized nation building in jeopardy. Darwin was not interested in creating a universal narrative that cast nature and human societies in analogous terms. More importantly, his notion of temporality was perhaps the most damaging aspect of his theory, because his account of evolutionary contingency undermined the notion of a universal timescale along which the shared progress of all individuals toward their unified destiny could be measured. Because populations evolved in different directions and at different speeds, both the unified destiny and the uniform temporality of progress were called into question. As was correctly understood not only in Argentina, but elsewhere in Latin America, the Darwinian revolution left a philosophical void that was difficult to fill.

The old Kantian idea of a philosophical system appealed greatly in Spanish America and continued to operate in the reception of Darwinism. In 1877, for example, the *Revista de Cuba* published an article by Julián Gassié (1850–1878) on the work of Haeckel that makes it clear how Haeckel could be read as having addressed the metaphysical objections of someone like Burmeister.[20] According to Gassié, since 1868, with the publication of Haeckel's *Natürliche Schöpfungsgeschichte* (Natural History of Creation), Darwinism had increasingly become the focus of revolutionary thought, such that by 1877 it boasted "hundreds of partisans on both continents" (Gassié 1877, 256). Germany's receptivity to Haeckel's Darwinism owes much, Gassié claims, to Goethe, a precursor to Darwin, to "the genial qualities of the [German] race, and to the preparation of its spirits by way of the well-known Hegelian principle of *process* (the metaphysical conception corresponding to the scientific doctrine of evolution), which greatly facilitated [Darwinism's] diffusion. What's more, thanks to the capacity for generalization and breadth of spirit that have made the Germans, like the Aryans of India, the synthetic race *par excellence*, the German savants have applied the principles of Darwin's system to linguistics, psychology, history, morality, medicine, and nearly every branch of the human sciences" (257).

Darwinism per se, with its implicit materialism and its emphasis on struggle, may have been too inconsistent with past ideas—but as corrected by Haeckel, supplemented by a metaphysical account of progress and reconciled with German idealism, it becomes much more congenial to the pre-Darwinian ideology that had been so important in Spanish America. Whereas Darwin lacks Haeckel's "generalizing intelligence," the German naturalist combines intelligence, imagination, and intuition, allowing him to "comprehend nature in its unity, without losing sight of it in the details of merely empirical research." Darwin offers no philosophical system in which his ideas might be reconciled with those of the past, nor is he interested in formulating the kind of synthesis that might serve as the underpinning of a well-defined philosophical movement. Gassié agrees with Haeckel that "empirical naturalists who don't take the trouble to arrange their observations philosophically, or who lack any general insight, do very little toward the advancement of science." The main worth of their "painstakingly collected details consists in the general results some more comprehensive intellect will extract from them later" (262).

The recognition of the philosophical problem brought on by Darwinism was not unique to Spanish American thinkers. French positivism, which was very influential in Argentina, was caught in the same dilemma. Emile Littré, who was responsible for the dissemination of Comte's ideas in France and England, regarded the new theory as "rationalistic and not experimental, ingenious but lacking sufficient evidence." More importantly, he considered the new evolutionism a matter "of purely biological and not of philosophical interest; even if Darwin's hypothesis were proved, it would carry no philosophical implications" (Simon 1963, 25–26).

Gassié's article is representative of a broad-based tendency on the part of Spanish American thinkers of his generation to look to Haeckel for a philosophical correction of Darwin. This tendency must have been galling for Burmeister, but it shows how, among intellectuals who labored under the Spanish colonial legacy, Haeckel's work could be read as an attempt to reconcile old and new ideas. Contrary to those who believe that the chief participants in this debate were unable to distinguish Darwin from

Spencer, for example, we are more inclined to read them as deliberately "correcting" both, in an attempt to produce a self-conscious evolutionary fusion that would prove beneficial to the country.[21]

In most cases, this blending of the Positive method with Darwin's, Haeckel's and Spencer's ideas reflects not ignorance, but a creative effort bent on overcoming the contradictory character of civilized thought. When an intellectual went with Spencer over Darwin, it is not because he failed to understand the latter's work, but because he thought, with some reason, that the political consequences of his analogies contrasted far too sharply with the faith in progress he felt was essential to the idea of civilization and the successful organization of the nation. Darwinism was a complicating factor in Positive Science's straightforward method for bridging scientific and political worlds. Its lack of any strong teleology undermined the inevitability of progress—and thus the obsession with progress so typical of the intellectual discourse of the 1880s is a sign of the perceived need to bolster a belief in it that the advent of Darwinism had called into question.

During the 1870s, the recognition of the revolutionary character of Darwin's theory is revealed by the interest among intellectuals in the self-conscious admixture of all sorts of evolutionary thought, derived from English, French, and German sources. As we have discussed in the case of Rawson, Lamarck's ideas were important well before Darwin, and with the publication of *Descent of Man* in 1871, quasi-Lamarckian approaches to inheritance made a comeback. In Argentina, as we have suggested, those interested in Darwinism were working to appropriate this revolutionary theory while simultaneously "correcting" it. And much the same was true elsewhere. As the work of Ernst Haeckel and later August Weismann demonstrates, there remained tremendous gaps in evolutionary theory, especially with regard to the mechanisms of heredity and variation, and consequently a great need for a new synthesis that would explain what, exactly, evolution was and how it worked. Most of the figures covered in this book were engaged in one or another aspect of the attempt at synthesis, though their approaches varied with their particular areas of expertise.

The tensions brought on by the Darwinian scientific revolution are nowhere more in evidence than in the later works of members of the Generation of 1837 who had been responsible for the supremacy of science in Argentina. Two of its most prominent intellectuals and politicians ended their lives trying to make sense of the new evolutionism, and how it fit or failed to fit their designs for the country. Juan B. Alberdi wrote his *Luz de día* (Daylight) while in London, publishing it in the same year as

Darwin's *Descent of Man*. In one section, he contemplates the consequences of applying Darwinian thought toward the organization of a nation. In a pointed satire, Alberdi observes that Darwinism has "heated" a great many heads, singling out Sarmiento for particular ridicule. Slyly, he remarks on the close association between the promotion of the new science and the consolidation of power in the hands of Bartolomé Mitre and his followers (Paz 1979, 69).

Sarmiento's considered response came in 1883, with the publication of *Conflicto y armonías de las razas en América* (Conflict and Harmony of Races in America), in which he updated his famous *Civilization and Barbarism* to reflect Darwinian insights, while also refuting the charges leveled against him by Alberdi and others. Both authors offer mixed interpretations of Darwin, revealing just how hard they were struggling toward new conceptions of progress and civilization (see Novoa 2007). Both of them, but particularly Sarmiento, wanted to add design to Darwin so as to allow them to conclude, as they always had before, that evolution was a force toward perfection in accordance with some rational plan. In short, they tried to assimilate the new ideas to the tradition they had been defending since their youth.[22]

It is difficult to pinpoint exactly when each of them first read Darwin's evolutionary work. Sarmiento wrote in 1868, while en route from the United States back to Buenos Aires, that he was aware of the disputes between Agassiz and Darwin, further asserting that "Darwin's theory was Argentine" and that he would "nationalize it with the help of Burmeister" (Sarmiento 1900j, 321).[23] He was evidently unaware of Burmeister's opposition to Darwin. Alberdi must have become aware of the new theory by around the same time, if not earlier. In 1869 he wrote that, if we accept a science that sees "the mixing" of races as the path to the perfection of the species, "racial distinctions make no sense." Lending further authority to this claim, he asserts that "the naturalist Darwin has cleared away any doubt concerning this natural truth, that is much more important than generally believed to the freedom of the human lineage [*género humano*]" (Alberdi 1899, 349). By the 1870s, Alberdi made use of Darwin in his political writings. He found the latter congenial in part because both he and Darwin had been influenced by the same sources in economic theory.[24]

During the decade of the 1870s, allusions to Darwin no longer consist mostly of references to his geological studies.[25] Now they are informed by a shared understanding of the scope of the revolution his discoveries had sparked. Many of the major intellectuals of the time had, in one way or another, weighed in on the new theory of evolution. In 1875 Eduardo Holmberg (1852–1937), then a medical student, wrote *Dos partidos en lucha* (Two Battling Parties), a fantastic allegory of the battle between those who favored the Darwinian revolution and those who opposed it. Holmberg's representation of the opponents is particularly interesting, as their motives include not only religious conservatism, but also adherence to traditional ideas on the meaning of civilization.

Holmberg's intervention coincided with the reorganization of the University of Buenos Aires in 1874 and the inauguration of its School of Physics and Natural Sciences the following year (Pyenson 1978). After the establishment of the the Academy of Sciences and the National Observatory in Cordoba, in 1869 and 1871 respectively, the Academy of Mathematical Sciences and the Academy of Physics and Natural Sciences (1874), the Museum of Paleontology and Archeology (1877), and the Geographical Institute (1879) all followed in quick succession (see Terán 1986). The capitalist expansion and the modernization that ensued had prepared the way for the introduction of new scientific ideas and especially for a revolutionary transformation in the understanding of humanity's relationship with nature. Also in 1875, Holmberg introduced Darwinian science to the curriculum at the Escuela Nacional de Maestras in the hopes of encouraging future teachers to bring the latest scientific developments into their own classrooms. This move was highly controversial, and Domingo Sarmiento emerged as the chief advocate for teaching Darwinism in the schools. Sarmiento's opposition to religious domination of education is evident in several pamphlets in which he opposes the teaching of creationism (Sarmiento 1900f, 92; orig. publ. 1881).

The importance that science had acquired in Argentina by the mid-1870s was also recognized outside the country. In March 1876, C. Gilbert Wheeler (1836–1912), professor of chemistry at the University of Chicago, dispatched a report from Buenos Aires on his experience in the country. Published in *Popular Science* the same year, his account offers an interesting picture of a scientific community in the midst of the upheaval of revolutionary change. Wheeler captures the tension between Burmeister and his colleagues when he refers to the "anomalous relation" between these scientists, subordinated to the Academy of Sciences and its president, and the members of the faculty at the University of Cordoba.

Wheeler also notes the devotion of certain young members of the educated elite to the study of science. In addition to a "considerable number of foreign eminent men," Argentina had a few "natives, mostly younger men, who are devoting themselves to scientific pursuits" (465). As we have noted, these young men were mostly part of the local elite that took research as a patriotic act, and began to chafe against Burmeister's authority. They gathered in the Sociedad Científica Argentina (Argentine Scientific Society), where they had a comfortable space "well supplied with scientific periodicals. There are seven hundred books in the library" (464). The society hosted a lecture every July 28th, and Wheeler makes a point of noting that at these events "ladies are also present" (464). Finally, the report ends by mentioning the scientific journals published at the time. Significantly, the author comments on the interest that the popular press had in publishing scientific communications. He explains that the prevalence of "some taste among the general public for scientific reading is exhibited by the circumstance that the daily papers find it worth their while to frequently admit scientific articles (467).

The Sociedad Científica Argentina was founded in 1872, its agenda largely dictated by government interest in harnessing science in the service of national identity. As Oscar Terán (1986) has argued, this interest was characteristic of Latin American (as opposed to European) positivism; Argentine positivism, in particular, was *programmatic*. Programmatic relationships between new scientific ideas and the modernization of the country are in evidence throughout the early volumes of the society's *Anales*. Naturalists' accounts appeared side-by-side with articles on economic modernization. The January 1877 issue featured a geological study, a statistical survey of the railways of the world, and a piece on indigenous customs, including the use of sulfur in treating animal hives.

Darwinism was thus catching on at a time when members of the intellectual elites saw science, industry, and economics as closely intertwined in the program of modernization. But uniting such diverse activities under the umbrella of the Sociedad Científica also caused new tensions. In a January 1876 meeting of the Sociedad, Estanislao Zeballos complained about the relative abundance of funds for the purchase of books on engineering, compared with the scarce allocation for books in other scientific fields. He argued for reapportionment. Luis A. Huergo, an engineer, disagreed, replying that, in his view, the Society's library was quite representative. Zeballos rejoined that it was absolutely "necessary to acquire for the library a collection of scientific works on the Republic, including those of D'Orbigny, Darwin, Bravard, etc." (1877, 9–10). His reference to Darwin is fairly typical of this period, by which time the Englishman was already generally acknowledged as one of the world's foremost scientific minds.

The fact that Darwin had studied Argentina placed the country in contact with a form of modernity that Zeballos sought to enhance. Argentina was part of the story of modern science, and so science must become part of the story of modern Argentina (Zeballos 1877).

In addition to its active members, the Sociedad Científica included such prestigious honorary members as Guillermo Rawson, Pedro Visca, Mario Isola, Hermann Burmeister, and the astronomer Benjamin Gould. It also had its international correspondents, including Leon Domecq in Spain, Pellegrino Strobel in Italy, Ladislao Netto, director of the Museum of Rio de Janeiro, in Brazil, and John Lubbock and Walter F. Reid in London. Lubbock, for his part, was close to Darwin, a frequent correspondent and sometime neighbor. Reid also knew Darwin, and it was he who communicated the Society's offer of honorary membership, which Darwin accepted in 1877.[26] In 1878 the Academia Nacional de Ciencias (National Academy of Sciences) in Córdoba followed suit with an honorary membership of its own, and in 1879 Darwin also accepted honorary membership in the Sociedad Zoológica (Zoological Society).[27] Darwin's election to the Sociedad Científica (Scientific Society) triggered a crisis that resulted in the resignation of Carlos Berg, a close ally of Hermann Burmeister's (Montserrat 2001, 13). The most important scientific association of the land had endorsed Darwinism, over Burmeister's objections.

This Darwinian turn by the scientific institutions coincided with the strengthening of ties between these same institutions and the government. In an 1877 letter to the Minister of Justice and Education, Guillermo White, president of the society, and Estanislao Zeballos, its secretary, reported their decision to fund an expedition by Ramón Lista to Patagonia (Lista 1879, 9). Lista was inspired by the widely reported success of an expedition by fellow Burmeister protégé Francisco P. Moreno a year earlier. In his application for funding, he touted potential benefits both commercial and anthropological, claiming that because "the origins of American man are lost in the mists of time," anyone interested in "reconstructing the customs, beliefs, and physiognomy of prehistoric races" would do well to look to Patagonia, where "in the bowels of the Earth the remains of an autochthonous dolicocephalic race" may yet be found (8).

The search not only for the origins of species, but also for the origins of Argentine man, would become a guiding theme, with clear political implications, for the homegrown naturalists who now explored the country. White's letter to the Minister of Justice and Education notes that such expeditions exert an "enlivening influence on the spirit of our youth, encouraging them to devote their ardor to significant research" (Lista 1879, 10). And like Lista, young people who dedicated themselves to naturalistic studies inevitably saw themselves as following in Darwin's footsteps. In Lista's words, Darwin, D'Orbigny, Bravard, and Burmeister had left "much to study, and not a little to discover, in this fantastic land where the poetic imagination of the Spaniards sited the city of the Emperors" (Lista 1879, 8).

The extent of the expansion of science during the 1860s and 1870s becomes even more apparent when we consider the number of scientific journals that had come into being by 1876. In 1879 the Harvard Library published its *Catalog of Scientific Serials, 1633–1876*. In its preface, Samuel H. Scuder notes that it has become impossible "for the working naturalist or physicist to keep track of the rapid growth of scientific literature during the past twenty years" (1879, viii). For Argentina, the *Catalog* lists eleven institutions and their journals, none of which had existed a mere fifteen years earlier (260).

The 1880s saw a renewed call for naturalists to ply their trade, especially in the recently pacified southern provinces. The "Desert Expedition" of 1879 served the purpose of heralding a new era in the emergence of modern Argentina. The forced removal of the indigenous populations and their supposed extinction after coming into contact with civilization were touted as a sign of evolution at work (Novoa 2009a). In their later recollection of the 1879 campaign, Adolfo Doering and Pablo Lorentz described it as the "final, definitive operation to subject the entire Southwest of the Argentine Republic to civilization and cultivation" (Doering 1916, 301). Concurrent and subsequent scientific expeditions were mounted to investigate natural conditions in the newly conquered territories and to properly certify the extinction of the Indians.

For its part, the Academy of Sciences was quick to recognize the scientific opportunities inherent in the Patagonian campaign. Oscar Doering, its interim president, wrote Julio Roca, the general in command of the expedition, to request his assistance in collecting zoological, botanical, and mineral samples. Doing so would, he claimed, "be a great service both to the country, and to science, simultaneously enriching the collections of national museums and bringing to light animal, plant, and mineral specimens the like of which may perhaps be found only in that part of the Pampa" (302). This letter triggered the organization of a scientific expedition operating in parallel with the military campaign. Its members were Adolfo Doering, a zoologist; Pablo G. Lorentz, a botanist; Gustavo

Niederlein, botanist's assistant; and Federico Schulz, zoologist's assistant. Their accounts serve to document not only their scientific findings, but the close relationship between scientific, military, and political projects of Argentine nation building.

As we mentioned, 1879 also saw the publication of the third volume of the French edition of Burmeister's physical description of the Argentine Republic, first published in German four years earlier (Burmeister 1875). The first volume of the French edition appeared in 1876 with an effusive dedication to Domingo Sarmiento, the author's "protector and most excellent friend" (1876). In testimony to the ongoing official sponsorship of science, and of Burmeister in particular, subsequent volumes were printed in Buenos Aires, appearing when treasury funds were made available to subsidize their publication. As discussed above, the third volume is of particular interest, as it reveals that by 1879, even Burmeister felt obliged to clarify (or revise) his views on Darwinism.

As we noted, in place of the stern objections of 1866 and 1870, he now offers a tepid endorsement (Burmeister 1879, 11–12). But Burmeister's close identification of the Darwinian hypothesis with the hypothesis of "original generation," a claim so ancient as to be traceable to pre-Socratic "philosophers of the Ionian school" (10), suggests that, like many of his Argentine contemporaries, he was unwilling to break with his traditional legacy. Even when forced to acknowledge the apparent truth of some of Darwin's hypotheses, he still insists on maintaining the old conceptions of progress, perfectibility, and design. Furthermore, as he makes clear, fundamental differences in type are what separate not only Linnaean phyla (insects and vertebrates), but species—"or to speak of man, black [from] white" (13). The conjunction of a quasi-Darwinian evolutionary hypotheses with a racial essentialism based on the existence of ideal types would continue to be important in Argentina for years to come and was one of the ways in which idealism managed to survive, blended with a strict Darwinian materialism. As we have noted, Haeckel provided the transition that Burmeister refused to consider. Racial archetypes, defended by both, would become the expression of the spiritual and aesthetic values of the nation, as we will discuss in the coming chapters.

CHAPTER THREE

The Triumph of Darwinism in Argentina

Darwinian Rule

We have touched on the close relationship between scientific exploration and military campaigns, including the extermination and forced displacement of indigenous groups in the late 1870s. This relationship was abetted in part by a new, quasi-Darwinian articulation of racial essentialism. From 1880 on, the Indians were no longer considered part of the contemporary Argentine population and were represented in scientific discourse only as tokens of an extinct past, of prior stages in the country's evolution.[1] Beginning in the 1870s, the collection of Indian remains had become a common way of commemorating the inevitability of their demise on coming into contact with civilization. In the 1880s, following the conquest of the southern regions, Moreno emerged as the naturalist most closely identified with the call to reconstruct the Argentine past, as a necessary condition for representing the evolution of the modern Argentine population. In an 1882 address to the Sociedad Científica, he outlines the origins of South American races and civilizations, asserting that "man did not simply appear like some meteorite descending from the sidereal realm. He has a long evolutionary pedigree" (Moreno 1882, 4).

Moreno continues his address, with frequent reference to Huxley, explaining the operation of evolutionary forces within the human species and the important task of the naturalist in reconstructing old lineages. He concludes by proposing that all the indigenous relics collected so far be assembled for a great anthropological exhibition in 1884. Its purpose would be to display to the contemporary Argentine "the long sequence of the physical and social evolution of his predecessors, from that humble primitive animal, physical man, who made no use of the intellectual spark inside his brain, all the way to the great, wise, conquering legislator, who raised the cities now scattered in ruins throughout the territory of the Republic" (43).

The fact that the popular press kept close track of news from various exploratory ventures, as well as hosting debates on evolution, demonstrates that the dissemination of Darwinism occurred at various levels of society. We agree with Thomas F. Glick that there is no fixed line between its scientific and popular reception (1989, 36). By 1880, references to Darwin could be found in virtually every newspaper and magazine, and it is likely that most of the literate public had formed some sort of opinion on the meaning of the new science. The publicly expressed views of political and scientific leaders responded in various ways to such opinion, and the close collaboration between political and scientific communities helped, in turn, to popularize Darwinian thought. High-profile exhibits were part of this popularization. In 1878 the Industrial Club set out to mount a South American Exhibition in commemoration of the tricentennial of the founding of Buenos Aires. The exhibition finally came about in 1882, and one of the centerpieces of the Argentine contribution was a collection of fossils assembled by Florentino Ameghino, included with the expressed purpose of "explaining Darwin's theory" to the viewing public (Sarmiento 1900e, 152).[2] More than the fossils themselves, what was on display was Argentina's identity as a modern nation, of which science and industry were essential components. In the absence of impressive industrial achievements, Ameghino's work thus occupied a critical place in this showcase, a paradoxical token of both the nation's antiquity and its modernity.

This exhibit garnered accolades from jurors, spectators, and reporters alike. Domingo Sarmiento himself paid homage to Darwin in the display hall, surrounded by Ameghino's fossils. His address begins by symbolically identifying the beginning of modern Argentina with Darwin's arrival aboard the *Beagle*. Placing his younger self at the site of this historic landing, he muses that Darwin may well have been among the officers he met on that occasion. Returning to the present, Sarmiento ruminates on the magnitude of the controversy sparked by Darwin's theory of evolution and speculates that the general public may still be unprepared to accept the new science. "The corollaries deduced from his theory cause great perturbation in the minds of those who cling to the old picture of creation, or successive creations" (1900e, 152).

In acknowledging the resistance of Christians and of those who held out for some notion of design, Sarmiento gives us a sense of how entrenched the divisions over Darwinism had become. He reports that even his friend Burmeister considered Darwin's view "a mere hypothesis, for it goes against the procedure of experimental science" (1900e, 152). Finally, he announces his intention to explain, in a future lecture, the connections between Darwin's evolutionary theory and his observations while in Argentina.

These were recurrent themes for Sarmiento throughout 1882, the year of Darwin's death, an event he commemorated with a lengthy eulogy at the Círculo Médico (Medical Circle). He also penned several pieces in the newspaper *El Nacional* on the specifically Argentine significance of Darwinian research, including the opportunity it presented for Argentina to assume a position of leadership in the scientific world. In a July 13 article on Ameghino, he congratulates the young paleontologist, a star among the "laborers who bring to light the treasures and arcana still hidden within the Earth," for having earned the jury's highest honors at the Continental Exhibition (1900a, 129). He returned to paleontology and archeology in another article the following day, noting that recent discoveries had shown that primitive man "had had a special theater in the Pampas and Patagonia in which to play out his development, or, as Sr. Moreno has it, the succession of types" (1900i, 127). Whereas the classification of European species was nearly complete, in the Americas the most crucial work was only now getting underway, with Argentines playing leading roles. Argentina's devotion to science and its successful assimilation of the new Darwinian science are, in his view, the best proof of the country's attainment of civilization. Sarmiento also alludes to Huxley, as he was wont to do during this period, as well as to himself, the absence of a byline allowing him to do so in the third person. "In explaining the law governing the succession of these prodigious and varied creations, Huxley prefers Darwin's theory, which, like its author, he calls the theory of evolution, on which animal species succeed each other, with earlier giving rise to later by means of gradual modification. Huxley, like Sr. Sarmiento, describes this process as evidence for intelligence, further asserting that if the theory didn't already exist, some pantheologist would have to invent it" (1900i, 127).

Sarmiento's general understanding of evolutionary theory was adequate, but he could not bring himself to accept that what Darwin was talking about was evolution without intelligence, without rational control, and without predictable outcome. His strong sense of historicism rejected the prospect of an unknowable future. For him to abandon the idea of progress that had been central to his plans ever since the struggle against Rosas would have been political suicide. Sarmiento revisited his interest in nationalizing Darwinism again in an August 30 article in *El Nacional*, reporting that Moreno was about to embark on another Patagonian expedition, in which he would "follow in Darwin's footsteps, continuing where he left off, or taking instruction from his insights on later discoveries. Darwin's observations on the primitive, fundamental role of Patagonia as an isolated center of creation have recently been confirmed in regions of North America that once occupied similar geological positions" (1900c, 177). And indeed, Moreno closely followed Darwin's account of his travels, frequently consulting it en route (*Anales de la Sociedad Científica Argentina* 1940, 281).[3]

In 1882 Pedro Alcacer was the president of the Círculo Médico, the association to which both Sarmiento and Eduardo Holmberg delivered their eulogies to Darwin the month after his death. In a lecture of his own, Alcacer, a Catholic leader, spoke against Darwin and his Argentine champions. Sarmiento replied in his August 30 piece in *El Nacional*, beginning by reiterating his assertion of a connection between the natural setting of Argentina and the theory of evolution itself. In his view, Darwin would never have made his great discovery were it not for his experiences in the Pampas. Argentines ought thus to embrace a system of ideas that so clearly expressed the destiny of the country. As he had promised as early as the 1860s, Sarmiento was, in effect, nationalizing evolution, for "Darwin plied his natural history in the fractured wastes of the Strait of Magellan, of the Galapagos Islands, and of the Argentine Pampas, scrutinizing tropical orchids, fossils embedded in rock . . . and the living animals of the Pampas" (Sarmiento 1900c, 185).

By the time of his death in 1882, Darwin was seen in Argentina as one of the founding figures of the country. This view was also reproduced in the popular press. In an article published in one of Buenos Aires's most important and respected newspapers, *La Nación*, shortly after the English naturalist's passing, Argentina is touted as "the first theater of his works" (*La Nación* 1882, 132). In this approach, accepted by most intellectuals by the late nineteenth century, Darwin and Argentina were inseparable. The naturalist had been able to transform science because of the research he had done in Argentina, and this research anticipated the country's destiny as a civilized nation with an important role to play in the development of scientific thought. According to this view, Argentine paleontological and geological evidence had helped Darwin to discover nature's secrets. It was on Argentine soil that he gained "the first revelation of the theory that made him immortal," having found there the "proofs of his system" (*La Nación* 1882, 133). As evidence in support of this thesis, the article quotes the first paragraph of *Origin*, in which Darwin expresses the importance of his observations in South America. In Darwin's original text, the passage reads, "When on board H.M.S. 'Beagle,' as naturalist, I was much struck with certain facts in the distribution of the inhabitants of South America, and in the geological relations of the present to the past inhabitants of that continent. These facts seemed to me to throw some light on the origin of species—that mystery of mysteries, as it has been called by one of our greatest philosophers" (C. Darwin 1859, 1).

Argentine intellectuals interpreted this paragraph as if South American nature had itself furnished the fundamental reasons and arguments behind Darwin's understanding of evolution. He had confirmed the importance of countries like Argentina for the study of science, and in doing so he had provided a mission for the young nation. Moreover, this relationship linked his name with the trajectory of Argentina's scientific progress. It was no coincidence, then, that he had died on the same day on which, nearly "half a century earlier, April 20 1834, Darwin set sail from the Argentine shore, at the utmost limits of its territory, for the last time." This connection is further reinforced when Darwin is quoted again, this time to describe the southern part of Argentina as "terra incognita," a description that in the author's view no longer applied, in part thanks to Darwin's labors. As if the author of *Origin* had waved a magic wand to give new life and purpose to the country, modern Argentina had been born together with evolutionary theory, the twin offspring of civilized thought and virgin South American wilderness.

> Today this *terra incognita* is the southern limit of Argentine territory, and now that its great explorer has departed the last shore of life to penetrate into the unknown reaches of death, his shade remains to guide the future settlers of that country, where in his honor, a mountain bears his glorious name.
>
> How the destinies of men and peoples are linked in the great chain of time and labor! How satisfying it is for a new nation such as our own to find its own name attached to the golden link of Darwin's, and to the genius and fruitful labors of that young Darwin of half a century ago! (*La Nación* 1882, 134)

Darwin was thus the link that bound the nation in the chain of civilized life. In 1885 Domingo Sarmiento gave a speech to celebrate the inauguration of the La Plata Museum, once again reinforcing this connection. Noticing that in the room with him there were hundreds of Indian skulls on prominent display, when at the same time Indians lived in the museum, he says that "it would be grounds for surprise in Europe" to learn that Argentina "has prehistoric men living in its territory." The purpose of linking the simultaneous continuity and discontinuity of Indian life in Argentina was to remind the people present at the celebration that the importance of this museum was "to collect before their disappearance the documents" of natural evolution. In line with his interest in linking science and aesthetics, he praises Francisco Moreno and Florentino Ameghino as "the young artists who decorated the [evolutionary] stage" (Sarmiento 1900i, 313).

If Sarmiento's end was to raise the profile of Argentina in the scientific world by raising the profile of evolutionary science in Argentina, then in several respects his goal was realized. The work of young Argentine naturalists gained international recognition, especially in France. Moreno's Patagonian expedition of 1876 was widely praised and his account of that voyage quoted in every book that touched on related subjects. In 1886, for example, Mathias Duval, a member of the French Academy of Medicine, published a collection of his lectures on anthropology under the title *Le Darwinisme*. Duval quotes Moreno, whose observations on matters evolutionary he takes to be authoritative. He praises Moreno especially for "contributing the discovery of intermediate forms" and notes his relationship with "his teacher and friend" Paul Broca (Duval 1886, 463–64). Eduardo Holmberg was another important figure involved in the classification of Argentine nature. His expeditions to Tandil during the years 1881 to 1883 were reported, for example, by Philip Lutley Sclater in his work on Argentine ornithology, published in 1890 in collaboration with William H. Hudson, who by then was living in England (Sclater and Hudson 1889, 226).[4]

Hudson was thirty-three years old by the time he arrived in England in 1874. He was known already for his correspondence with the London Zoological Society, which had been published in its proceedings in 1870. Part of the importance of these letters was that he attacked Darwin's assertions about Argentine birds in *Origin* in a way that cast doubt on Darwin's powers of observation. Juan José Parodiz has observed that this criticism "had set them apart. The older man was entirely disinterested in keeping in touch with the younger" (see Parodiz 1981, 91).

The exchange between the two is interesting because it is the only one in which an Argentine citizen argued directly and publically with Darwin himself. The first letter was written in Buenos Aires in December 1869.

At the time, Hudson was living in a house close to Burmeister's museum, and we may well imagine that this correspondence was not unknown to the Prussian savant. The second letter, written a few days later, criticizes the "unfortunate" assertion in the *Origin* that the woodpecker "never climbs a tree" (Hudson 1870a, 112). The third letter, sent around the same time, takes more care in attacking Darwin's theory and his argument in its support. It also points out the important role that observing nature in the Pampas had had in the crafting of this theory. With regard to the woodpecker, he explains the "erroneous account of its habits in Mr. Darwin's work, which makes it worthy of particular attention." According to Hudson, it was *the English naturalist's reliance on analogies* that was to blame for such mistakes. He denounces this kind of observation in the name of a more descriptive, detail-oriented approach that was not at the service of proving any hypothesis.

> In Chapter VI of his well-known work on this subject [*Origin of Species*] the author speaks of the altered habits, caused by change of habitat and other extraneous circumstances, and infers that it would be an easy matter for natural selection to step in and alter an animal's structure so as to make a new species of it, after its habits have been so altered. He then proceeds to ask whether "there can be a more striking instance of adaptation given than that of a Woodpecker for climbing trees and for seizing the insects in the chincks of the bark"; and, in reference to this, states that there is a Woodpecker inhabiting the plains of La Plata, "where not a tree grows," and which is consequently a "Woodpecker which never climbs a tree" (*Origin of Species*, 4th ed., chap. 6, pp. 212, 213).
>
> The perusal of the passage quoted by one acquainted with the bird referred to and its habits might induce him to believe that the author had purposely wrested the truth in order to prove his theory, but as Mr. Darwin's "Researches" were written long before the theory of natural selection was conceived, and abound in similar misstatements when treating of this country, the error must be attributed to other causes. (Hudson 1870b, 159)

Darwin replied a few months later with an explanation of his observations and the context in which he had made them. First, he acknowledges the authority of Hudson's work on ornithology. Second, he moves to defend his views, citing Félix de Azara and explaining that his observations were accurate for the region in which he had conducted them. Finally, he answers Hudson's veiled aspersions on the honesty of his work. "Finally, I trust that Mr. Hudson is mistaken when he says that anyone acquainted with the habits of this bird might be induced to believe that I "had purposely wrested the truth in order to prove" my theory. He exonerated me from this charge; but I should be loath to think that there are many naturalists who, without any evidence, would accuse a fellow worker of telling a deliberate falsehood to prove his theory" (C. Darwin 1870, 706). Hudson's memoirs acknowledge that his own issues with Darwin began shortly after he learned of his theory in Argentina in the early 1860s. According to Hudson, upon reading a copy of *Origin* brought back shortly after its publication by a brother who had traveled overseas, his reaction was to brand the theory a collection of lies. Interestingly, he explains that this outburst was motivated by Darwin's attack on the conception of a nature built in accordance with a benevolent design. "Thus the old vexed question—How to reconcile these facts with the idea of a beneficent Being who designed it all—did not come to me from reading nor from teachers, since I had none, but was thrust upon me by nature itself." He adds that he was able "to resist its teachings [those of *Origin*] for years, solely because I could not endure to part with a philosophy of life" that could not "logically be held" if Darwin was right (Hudson 1919, 214–15). Hudson's reluctance to accept the new evolutionary theory turned on a problem that affected most of the intellectuals of his generation: its destruction of the old ideals of harmony, beauty, and cooperation in nature.

As we have noted, this same philosophical problem helps to explain why, by the 1880s, we find the reception of Darwinism increasingly influenced by Ernst Haeckel. Dora Barrancos has argued that the impact of Haeckel's *Natural History of Creation* was so great that in Argentina "many readers were first introduced to Darwin's ideas by way of Haeckel, whose works served the dual purpose of offering new scientific evidence in support of transformism, while also popularizing the doctrine" (1992, 13). An aspect of Haeckel's writings that was particularly important was the connection he drew between scientific and aesthetic normativity. As Milulás Teich and Roy Porter have argued, the end of the nineteenth century saw "a challenge to aesthetics posed by the nineteenth-century growth of science and its institutionalization" (Teich and Porter 1990, 8). For those who felt that the new science was a threat to the old values grounded in the spiritual concern with culture and aesthetics, Haeckel presented a plausible alternative. The harmonious picture implicit in his nonreductive monism made possible the representation of beauty in nature, against the ugliness and chaos that analogies driven by the struggle for survival provoked. As Kurt Bayertz has shown, Haeckel restored a sense of both harmony and continuity in nature that skirted the more threatening aspects of a reductive materialism.

> The universe formed an evolutionary whole in which all the higher forms were presaged in the lower ones, so that nowhere were there any discontinuities to be found. Haeckel argues in particular against the assumption of a discontinuity in the emergence of mind and spirit from pure matter. For him "monism" means, above all, the recognition that even the highest intellectual capabilities of man had their roots in the instincts of animals, in the sensitivity to stimulation of the plants and the reactivity of micro-organisms. For him even individual cells have a "soul" (*Seele*)—and, since it isn't possible for the souls to have arisen out of nothing, he was convinced that the whole of inanimate Nature possessed spiritual qualities.... The Universe—both the animate and the inanimate part—is by no means dead and purely material, but living and filled with soul. (Bayertz 1991, 286)

This aspect of Haeckel's influence can be easily recognized in the importance that idealist morphology had in Argentina. In the previous chapter we mentioned how Sarmiento related the replacement of the *vaca ñata* with evolution toward increasingly perfect types, applying the same logic to all species. According to this analysis, evolution consisted in the pursuit of a perfect form representing national values. The corresponding concept of nation was spiritual rather than material, based not on the physical reality of the population, but on the aspiration of the spirit. But this aspiration led to the privileged position of certain races over others. This, in turn, entailed the presumptive invisibility of those considered divergent from the ideal national form. As witnessed by Sarmiento, this nationalization of morphology affected all species. To take another example, a popular manual for Argentine ranchers, written by Godofredo Daireaux, defends the same logic.[5]

> In the first years of the formation of a breeding herd, the timely elimination of the most inferior animals among both first and later generations must be the whole basis of the operation.
>
> Later on, if the growth of the herd exceeds the culling, as is generally the case, one must then exercise, upon this much larger herd, the perfect selection of those that most approximate the ideal type that, according to Darwin, every breeder must have clearly imagined and impressed upon his memory before undertaking to improve a breed. (G. Daireaux 1887, 363)

As suggested in this manual for breeders, the conception of an ideal type is a necessary precondition of the process of selection, an idea that Ernst Haeckel seemed to endorse. As Levins and Lewontin explain, Darwin had rejected Platonic-Aristotelian idealism and thus reoriented the understanding of evolution. "He replaced the ideal entities, species, with material entities, individuals and populations as the proper objects of study" (Levins and Lewontin 1985, 31). But this important shift was extremely problematic for the nation-building process, because it was precisely in the ideal form to be attained in the future that the present heterogeneous reality of the population might be mitigated. In resting evolution on the concrete reality of the bodies that the elites sought to keep invisible, Darwin dispelled the illusion on which the country's future had been predicated. As we will discuss in regard to sexual selection, beauty became the category that bridged pre- and post-Darwinian culture.

By the 1880s references to Darwinian and quasi-Darwinian conceptions of evolution had become ubiquitous, even making their way into the official report of the Buenos Aires municipal census. For census director Francisco Latzina, these conceptions were anticipated by Hippocrates's injunction that physicians study the seasons, weather, and soil, so as to gain "a precise notion of the particular diseases" associated with given environmental conditions. Similarly, "the views of the present day, the great breakthroughs of this century that have immortalized the names of Lamarck and Darwin, subordinate the customs, sentiments, genius, and even the political conditions of peoples to the influence of the climes in which they dwell" (Latzina 1889, 268).

Darwin's ideas also took hold in the naturalist literature of the 1880s, inspired by French authors (Grutzmann 1998). Hippolyte Taine and Emile Zola provided models for the literary assimilation of Darwinian themes. Literature also furnished an extrascientific standpoint from which to address the darker side of science. Anxieties surrounding the struggle for existence and the possibility of extinction, fears for which there was little room in strictly scientific publications, found voice in fiction. In one way or another, most leading scientists were also driven to literary pursuits, Carlos Octavio Bunge and Eduardo Holmberg being, perhaps, the most notable examples. Argentine naturalism was no mere imitation of the French model; where the French texts enjoyed success in Argentina, it was because they spoke to the Argentine scientific imperatives. The particular social preoccupations of French naturalism and their characteristic solutions are generally absent in its Argentine expression, which is pervaded by pessimism and fatalism. The growing popularity of this sort of naturalism demonstrates the extent to which evolutionary logic had permeated nearly every niche of Argentine culture. By the end of the century, the Darwinian perspective, moderated by residual positivist faith in political progress, had come to dominate in science, literature, and education alike.

But it did not pass unchallenged. Catholic resistance to consolidation of power in the hands of the *naturalistas* was strong and became particularly fierce in response to reforms designed to secularize civil marriage and public education. Once again, José Manuel de Estrada was a leading voice of the Catholic opposition. He specifically denounced the alliance between liberals and Darwinians, several times mentioning Darwin by name in an 1878 address before the Catholic Association of Buenos Aires. Argentina, Estrada charged, was in danger of going the way of Europe, where it was all too easy to "predict the removal of every moral brake on human behavior, leaving governments free, in their cynical materialism, to elevate force above right." In Estrada's view, the German Empire already embodied this trend, its society thoroughly subverted by "the chief weapon of *Kulturkampf*: the doctrine that converts an entire nation into an army, as blind and powerful as a machine, and as submissive as any herd; the hateful doctrine of those who trace man's genealogy to the apes, and consign his destiny to the law of the jungle" (J. Estrada 1905c, 163; italicized German text in original).

Estrada's understanding of the dangers inherent in the political and social application of Darwinism was perceptive and in some respects prophetic. His approach was not antiscientific; on the contrary, he continued to defend science itself, while warning against adopting, in the name of science, any policy that undermined the moral framework on which the good of the individual was subordinate to the greater good of society as a whole. He also criticized the flawed synthesis of Darwinism and positivism, which he saw as a characteristic defect of Argentine materialism. The threat of Darwinism consisted, in part, in the fact that it rested on arguments from analogy, where reasoning "based on generic ideas quickly leads, in the absence of particular judgments and precise definitions, to fallacious conclusions, and to doctrines as vague as any figment of the imagination" (J. Estrada 1905c, 165).

For our purposes, Estrada is a particularly interesting figure because, rather than simply defending religion from science, he sought to reconcile the two. One of his most celebrated polemics against Darwinism and associated liberal policies begins by acknowledging that, over the course of the nineteenth century, "man . . . [had] delved deeply into the bosom of nature, investigating previously unknown laws, and discovering marvelous applications." Darwin and other natural historians deserve praise for their "rare observational talents," despite the "villainous consequences" of their theories. But Estrada has nothing but disdain for the "indecent hoaxes of Haecke [sic]," the German champion of Darwinism. Science is a worthy pursuit, so long as it respects its own limitations; it can never "discover the higher bonds that tie it, under the metaphysical conditions of production, to universal harmony, subject to supreme law" (J. Estrada 1905b, 212).

Estrada correctly discerned that Darwin's explanatory model removed humanity from the center of a designed creation. What he called "naturalism" (*naturalismo*) was thus incapable of addressing the key existential question, "Where am I going?" (¿a dónde voy?) (1905b, 214). Its moral failure was a direct consequence of its refusal to elevate humans above the rest of nature. Furthermore, a society dominated by the close association of this brand of materialist science with industry was one doomed to stagnation. Anticipating Uruguayan writer José Rodó's critique of utilitarianism in his 1900 work *Ariel*, Estrada denounced the new conceptions of civilization for their betrayal of what it meant to be civilized: "Nineteenth-century society [has been] molded by naturalism, made prosperous by the physical and natural sciences, and proud of its industry. It is ruled by political economy, under the tenacious inspiration of Adam Smith, predecessor to Macleod and the first of the long lineage of sophists that gave rise to Bentham and Franklin, the patriarchs of utility, the great tacticians of calculated virtue" (1905b, 216).

Under Estrada's leadership, the Catholic opposition did not attack science so much as the social and political consequences of recent applications of science. Estrada predicted that once God had been abandoned in favor of the consensus that "material well-being is the supreme end of man, success in the Darwinian struggle for life [would become] the touchstone of justice." Any triumph could then be legitimized for having been "an expression of superior force." This morally decadent society would be ripe for such diseases as socialism and demagoguery (1905b, 218). The growing power of the state and its increasing control over social life were symptoms of this trend.

In Estrada's energetic closing address to the Catholic Conference of 1884, he once again denounced the naturalistic worldview that had taken hold of Argentina and called for a moral renewal to overturn it, restoring the nation's values. "Liberalism," he charged, "has subordinated men to the whims of parties and tyrants, themselves fatally determined by the laws of force and matter that give rise to all life, all society, and everything else we see." The naturalist's interests acknowledged "only that which weighs in his scales or is distilled in his alembic. . . . Darwin, Spencer, and Hubner . . . are his prophets" (1905a, 416).

While Estrada was sometimes prophetic, neither he nor many of his contemporaries on either side of public debates were particularly careful about which views they attributed to Darwin, which to Spencer, Haeckel, Huxley, or others. Lucio Mansilla (1831–1913) was clearly aware of this problem, addressing it in an installment of a weekly newspaper column published throughout the 1880s and 1890s. "The belief that Darwinism and evolution are the same thing is a confusion," he notes. "They are not. One must distinguish between the doctrines of descent with modification, and natural selection" (Mansilla 1963, 555–56).[6] Furthermore, it is important to acknowledge that evolutionary thought, now so closely identified with Darwin, has a significant prior history. This point made, Mansilla takes up one of the burning issues of the day, the question as to the causes of variation and the influence of the environment in shaping future generations. While refusing to declare his allegiance to any particular school of evolutionary thought, he stresses the significance of environmental influences, following the French tradition that had been so important in Argentina. "The importance I place on the environment or *entourage* is so great that, when I was among the Indians, obliged to do as they do—for, as it is said, 'He who lives among wolves must learn to howl with them, or be devoured'—I became convinced that color and scent, forms and mores, interior and exterior, may all come to be modified in far less time than one might expect, all by virtue of nutrition, habits of life, and examples and ideas found ready to hand" (558). As further evidence for the importance of the environment, Mansilla cites the case of two children who, orphaned in a massacre in the province of La Rioja, were discovered years later living like animals. They "spoke no language, communicating with gestures and a few vocalizations." Though themselves the offspring of modern humans, they appeared greatly to "resemble the first ancestors of man, just as Darwin describes them" (561). Evolution was an indisputable fact, but the relationship between modern, civilized humans on the one hand, and barbarians, members of inferior races, and the weak on the other, remained murky. As Mansilla was aware, his own preoccupation with the relative power of nature and nurture was widely shared throughout the 1880s. He considered his own observations important enough to merit sharing them with Darwin himself.

> Transformation and evolution may be both ascending and descending, progressive and regressive. And so we should not be so ashamed for having descended from an animal much like the apes, for in La Rioja we have seen men less intelligent than they, reduced to brutes. . . . On his deathbed, Charles Darwin, the great naturalist and author of *On the Origin of Species*, received a letter from me, dispatched from Rome, in which I recounted this case in detail as further confirmation of the theory of evolution, in the sense both of progress and its opposite, that he might add it to his vast store of observations. (561)[7]

If Mansilla appears indecisive about the range of evolutionary events consistent with Darwinism, we should recall that he was writing at a time when those who had assimilated *Origin* were still coming to terms with *Descent of Man*, published in 1871. Darwin's new discussion of sexual selection and his more nuanced stand on the role of the "soft" inheritance of acquired traits left the door open to those who sought to introduce Lamarckian and positivist theses into Darwinian orthodoxy. Sarmiento's 1882 eulogy for Darwin, in which he justifies a soft inheritance thesis on the basis of sexual selection, exemplifies this approach (see chap. 5). For as George Levine has argued, "Darwin's theory of sexual selection puts back into the theory of evolution the intention and cultural direction that seemed so alarmingly absent in Darwin's first formulation in *On the Origin of Species by Means of Natural Selection*. . . . Intention, that central motif of natural theology, went out with the *Origin*, only to return with *The Descent*" (2003, 37). *Descent* also complicated the already fraught discussions of gender in general, and feminization in particular, by making female mate choice a driving force in evolution in his explanation of sexual selection.

Darwinian Paths

The decade of the 1890s began with the inaugural issue of the *Revista Argentina de historia natural*, under the direction of Florentino Ameghino, by this time an established paleontologist and evolutionary theorist.[8] Published in 1891, it included an address delivered by Ameghino to the Instituto Geográfico Argentino in 1889, in which he reflected on the present status of Darwinian thinking. Not long ago, he claims, "to talk of transformism, or the evolution of the animal kingdom, was considered far too serious a task, one to be approached with excessive caution, so as not to collide with the deeply rooted ideas sown by conventional education" (Ameghino 1891, 18). Defending a stricter Darwininism than Mansilla, Ameghino is at pains to divorce himself from commonly misattributed non-Darwinian theses and to confront some of the more challenging consequences of Darwinism head-on. "No: Humanity has never regressed. It has advanced in accordance with the immutable laws of evolution. These laws are not fatal, despite what is often said to the contrary, but wise, because they are laws of nature, and nature is never wrong. As they push us ever forward, so do we always advance, though we litter the road behind us with waste of our remains" (18). Claims such as this have led some authorities, including Fernando Márquez Miranda and Julio Orione, to argue that "Ameghino only accepted Darwinian concepts in a general way and . . . was strongly influenced by Lamarck" (Orione 1987, 447; see also Márquez Miranda 1951; 1957). But when we recall that Darwin had said something very similar in *Origin*, we can see that even the English naturalist had shared to some degree a faith in perfectability. According to Darwin, "As natural selection works solely by and for the good of each being, all corporal and mental endowments will tend to progress towards perfection." But what is different is that in Darwin the prospect of extinction is mitigated by an optimism that springs from his very different cultural situation. In fact, having commented on the terminality of existence, he immediately reminds the reader that "we may look with some confidence to a secure future" of great length (C. Darwin 1859, 489).

In Argentina, Darwinian reflections on such notions as atavism, regression, natural selection, and extinction could not be mollified by a sense of optimism. Quite the contrary, fear and anxiety dominated these discussions. Darwin was read as a much more pessimistic thinker in a world in which his analogies operated differently than in his own cultural environment. Because of the analogical way in which biology applied to the study of society, on reading the end of *Origin*, one might well ask how the idea of nation, predicated on the continuity and permanence of its people, could be reconciled with the instability inherent in Darwinian principles. "Judging from the past, we may safely infer that not one single species will transmit its unaltered likeness to a distant futurity. And of the species now living very few will transmit progeny of any kind to a far distant futurity; for the manner in which all organic beings are grouped, shows that the greater number of species in each genus, and all the species in many genera, have left no descendants, but have become utterly extinct" (C. Darwin 1859, 489).

We believe that it is precisely in the context of this contradiction between national and scientific interests that Ameghino must be understood. In Ameghino's case, Lamarck's influence is not tied to the scientific defense of either the inheritance of acquired traits or realism with respect to the concept of species, both of which were common expressions of neo-Lamarckian predilections. He was not close to Lamarck the scientific thinker, but to the social ideas that spread under the umbrella term of Lamarckianism by the end of the nineteenth century. And this type of thinking was related not with scientific Lamarckianism, but with the teleology of nation, with guaranteeing that Argentina and Argentines had a place in the future evolution of humankind. In our view, the Lamarckianism that some twentieth-century scholars have attributed to Ameghino is the result of the changed perceptions that emerged from the debate between neo-Darwinians and neo-Lamarckians and the victory of the former by the early twentieth century. A full defense of this reading would take us well beyond the scope of this volume. For the present, it will suffice to recall that extinction is a constitutive, ineliminable aspect not of Lamarck's evolutionary theory but of Darwin's. Extinction, in turn, is the cornerstone of the evolutionary dynamics of Ameghino's phylogeny. We must recall that, through the end of the nineteenth century, there *was no* strictly Darwinian theory of inheritance and that Ameghino himself was not especially concerned with inheritance per se. As a paleontologist interested in phylogenetically driven taxonomy, Ameghino's primary focus was not inheritance but *variation*, a phenomenon on which Darwin had a great deal more to say than Lamarck. Extinction plays a constitutive role in Ameghino's concept of evolution, and this does not come from Lamarck, for whom evolution operates *without* extinction.

At any rate, according to Ameghino, by 1890 the need to tread lightly around matters evolutionary has passed, for "it is no longer necessary to waste time trying to prove a theory that has been adopted by all naturalists." In a characteristic swipe at his enemy Burmeister, Ameghino ridicules the venerable naturalist as the last remaining holdout of the old guard, to whose "wholly sentimental objections no one now feels obliged to reply."[9] Instead, present-day "naturalists are left free to reconstruct the great tree of life by tracing the lines of descent among species, a slow, patient task that demands many and varied investigations" (Ameghino 1891,18).

Though dismissed by Ameghino, Burmeister remained a figure of great renown in the scientific world, continuing to maintain his valuable contacts with scientists, journals, and other institutions. In 1890, for example,

Nature published a review of Burmeister's recent report on the collections of the Museo Público. The article leaves little doubt as to the reviewer's familiarity with and respect for Burmeister's work, praising him as "a veteran man of science" who has continued to publish the "'Annals' of the Museum under his charge with unfailing regularity." Burmeister's report also "gives us an account of a scientific expedition into Patagonia, recently carried out by his son, Sr. Carlos V. Burmeister, one of the assistant naturalists of the museum" (Annals of the Museum 1890, 293).

This expedition, charged with both geological and anthropological research, was undertaken in 1888. Julio Roca's previous military campaign must have been a success, for, as the *Nature* reviewer recounts, the younger Burmeister discovered that much of Patagonia "appears to be almost deserted at the present time. No natives seem to have been met with between the Chubut and the Rico Chico de Santa Cruz until the lower part of that river was reached." For geologists, photographs taken during the expedition "will be of interest in connection with the question of the singular 'basaltic terraces' of this country, of which Darwin gave us the first indication, and which are frequently referred to by Señor Burmeister" (Annals of the Museum 1890, 294). Whatever Burmeister's views on the subject of evolution—and we have argued that they were far less clear than Ameghino seems to have believed—his high regard for Darwin the geologist remains unquestionable, as does the high regard with which the global scientific community continued to hold Burmeister.

Ameghino's Darwinian triumphalism also served as a rhetorical tool in his dismissal of Haeckel's ontogenetic approach to evolutionary phylogeny, in favor of his own phylogenetic method.[10] Ontogeny could furnish only "the broadest outlines of evolution, obscuring most of the details," while Ameghino's phylogeny, which employed "a calculus for discovering all of the forms ancestral to a given species," was better suited to "descending into the most minute level of detail" (Ameghino 1891, 18). It is in his interest in variation and individual difference that Ameghino's Darwinism is most evident. His resistance not only to embryology, but to other sorts of developmental evidence was so thorough that, as Simpson (1984) reports, he tended to classify a pair of specimens that any other paleontologist would have identified as juvenile and adult members of the *same* species as belonging to two entirely *different* species. His prejudice in favor of a strictly morphometric, quantitatively precise method is in part explicable as a response to the scruples of Burmeister's generation, who continued to lament the absence of any inductive method in Darwinian evolutionary theory. In consequence of this peculiarity, most of Ameghino's prodigiously named species are no longer recognized. But as recently as the completion of this book, John Alroy of the National Center for Ecological Analysis and Synthesis reported at least 320 Ameghino genus names still current in their Paleobiology Database, making Ameghino in this respect the most productive mammal paleontologist of all time, well ahead of runner-up Edward Drinker Cope (John Alroy, personal communication).[11]

In the waning decades of the nineteenth century, the association between evolution and extinction became increasingly close, and, in consequence, progress in the evolution of the nation was best demonstrated by evidence for the extinction of those less favored by nature (see chap. 4). If before Darwin civilization depended on the creation of a culture representing ideal values, now it was related to the creation of a race representing an ideal type. As Moreno affirmed, the future Argentina would contain a necropolis, populated by those who had been eliminated in the evolution of the modern nation, finally in sync with the most advanced nations of the world. "As early as 1878, I painted in broad strokes a picture of the great confusion of the races of South America, and the presence in Patagonia of the remains of men who had emigrated from the north of the Continent. Human crania, the residue of industry, and the inscriptions on cliffs, all prove this. The Argentine Republic is without a doubt a vast necropolis of lost races. They came from the remotest theaters, pressed on by the fatal struggle for life, in which the strongest always wins, and here in our southern extremes, the conquerors annihilated the conquered" (Moreno 1890–1891, 50).

If what had happened to the indigenous populations was the expression of a law of nature, it was fated to happen again. During the decade of the 1890s, the psychological condition of a population whose members perceived themselves as belonging to groups less fit than others for the struggle for existence came under increasing scrutiny. Darwin's insights provided a new perspective on the relationship between biology and behavior, because they undermined the presumption of a teleological connection between emotions and their physical manifestations. Behavior could no longer be taken as exclusively the expression of intelligent agency (Ghiselin 1973, 965). In Argentina, Carlos Octavio Bunge took up the psychological consequences of these discoveries.

By the 1890s, naturalists, writers, scientists, and educators alike were deeply influenced by the new evolutionary ideas. Julia Rodríguez has observed that by this time "Argentina emerged as an up-and-coming and promising station in the world of transatlantic ideas. The Argentine state took pains to promote its investments in science by sending ambassadors and reams of published research abroad" (Rodriguez 2006, 29). One of the most important cultural exchanges of this period involved not England, France, or Germany, but Italy. During this decade, the Italian school of criminology became important through the work of Cesare Lombroso and his disciple Enrico Ferri (Pavarini 1983). The influence of this school was most apparent among positivists and socialists, who recapitulated the same split that had taken place in Italy (Lombroso, Ferri, and Garofalo 1886). Lombroso wanted to apply the positivist method to the understanding of crime, but in his organic approach to the development of the criminal mind, he turned to Darwin. His *L'uomo delinquente in rapporto all' antropologia, alla giurisprudenza ed alle discipline cacerarie* (The Criminal in Relation to Anthropology, Jurisprudence, and Penal Discipline) was published in 1876, becoming a standard work for advocates of a new approach to criminality and law (Lombroso 1876).

Darwin was particularly important in providing a foundation for Lombroso's account of the evolutionary basis of criminality, as is evident in his famous analysis of criminal women (Rafter and Gibson 2004). Like his Argentine contemporaries, Lombroso was engaged in a reassessment of the culture of the Enlightenment after Darwinism. His work tried to bring together politics and science, and it was connected with these particular contexts: "the unification of Italy; the growing prestige of science, specifically Darwinism; the revolt against Enlightenment legal theories; and the birth of sexology" (Rafter and Gibson 2004, 15). His attack on the ideas of legal egalitarianism that were popular in Italy was appealing to Argentine intellectuals seeking new approaches to understanding society in the wake of the first responses to Darwin. Because Lombroso's ideas drew on a combination of positivism and evolutionism that was already well established in the region, his work quickly gained traction. In 1888 Lombroso himself acknowledged his newfound importance in the Western Hemisphere in his prologue to the Italian edition of Argentine lawyer Luis María Drago's *Hombres de Presa* (Predator Men) (Lombroso 1890).

The other relevant source in the diffusion of the Italian criminological school was the work of Enrico Ferri. Unlike Lombroso, Ferri tried to bridge socialism and Darwinism using the positive method characteristic of this school. In the earliest period of his development, Ferri "is almost wholly under the influence of Spencer's sociology. At a later period, however, he became a convert to the practical program of Marxian socialism" (Ellwood 1917, xxix).[12] In his view, "Marx complements Darwin and Spencer, and together they form the great scientific trinity of the nineteenth century" (xxviii). In Argentina, socialism would play an important role in the continued spread of Darwinism. Like the Argentine socialists, Ferri saw himself as working in the spirit of the "modern scientific revolution" that had triumphed in his times "thanks to the works of Charles Darwin and Herbert Spencer." This was an attempt to make Darwinism a seminal critical theory that could reach fruition with the incorporation of Marxism. Ferri recognized that Darwin and especially Spencer halted "when they had traveled only half way toward the conclusions of a religious, political and social order, which necessarily flow from their indisputable premises." He saw Marx as the one who completed "the renovation of modern scientific thought" (Ferri [1894] 1900, 10). The introduction of Marxist ideas solved an old problem that had obsessed Argentine intellectuals during the previous decades, the problem of the role of progress and design in nature. A synthesis organized around Marxian ideas allowed for the restoration of the traditional sense of vertical progress in civilization.

The fear that following Darwinism would lead to socialism had been often expressed in Europe, particularly in Germany. It was there that, in 1877, a famous debate took place between Ernst Haeckel and the pathologist Rudolf Virchow. The latter declared that Darwin's ideas led directly to socialism. Haeckel responded "by pointing out that Darwinism was 'aristocratic' rather than democratic, let alone socialist, because of its principle of the survival of the fittest which had been translated into 'victory of the best'" (Maasen and Weingart 2000, 49). In Argentina, Haeckel's ideas were widely used by socialists and antisocialists alike. As in Germany and in other parts of Europe, the use of Darwinism as an ideology "is not limited to one group whose interests it supposedly matches, but rather by many groups who interpret as they choose." This usage "both ubiquitous and heterogenous, renders it problematic, if not outright false, to assume a specific impact of the theory," either instrumental or legitimating. But the multiple uses of new evolutionary ideas point to the fact that all the participants felt that they could not make a point without a proper explanation based on science. As elsewhere, the dominance of scientific thinking "enforces the usages of (that is, Darwinian) terms and concepts, which, in the course of this happening, assume their (at times bewildering) connectivity to a multitude of discursive areas" (Maasen and Weingart 2000, 49).

Members of the elites had recourse to Haeckel's aristocratic interpretation as a proof that Darwinism was right. Socialists, such as Ingenieros, also used Haeckel, as well as Ferri's approach, to prove that Darwin was correct and that, with the benefit of Ferri's synthesis, socialism could complete the scientific revolution that would give rise to a better society. According to Ferri, socialism would guarantee that natural law operated unopposed, abolishing the protection of weak individuals who survived on class privilege alone. As we have previously discussed, the earlier synthesis between positivism and Darwinism was an attempt to provide a sense of progress and purpose that could be reconciled with the natural laws of evolution. The synthesis between Darwinism and socialism tended to do the same, to guarantee that the human will could control the course of evolution and that society's improvement could be assured. Both attempts at synthesis introduce a teleology geared toward dispelling the worry that the appearance of progress was merely the result of chance events. Ferri's rejoinder to Haeckel's aristocratic interpretation makes this clear. "I have already answered this objection in part by pointing out that socialism will assure to all individuals—instead of as at present only to a privileged few or to society's heroes—freedom to assert and develop their own individualities. Then in truth the result of the struggle for existence will be the survival of the best and it is for the very reason that in a wholesome environment the victory is won by the healthiest individuals. Social Darwinism, then, as a continuation and complement of natural (biological) Darwinism, will result in the selection of the best" (Ferri 1900, 56).

As we have noted, this desire to turn Darwinism into a comprehensive theory that could be connected with every aspect of human life was, in part, an expression of the desire to maintain continuity with a cultural tradition that preceded Darwin. It was important to force the theory of evolution to conclude in a categorical way that humanity's privileged place in the universe would continue and that the future would consist in continuous progress toward humankind's eventual perfection. The restoration of a strong sense of teleology and historicity not always present in Darwin's writings was typical of most Darwinists, both the socialists and their rivals.[13]

At the end of the nineteenth century and the beginning of the twentieth, several new publications arose to promote and explore the new Italian ideas on criminality. Luis Drago, Norberto Piñero, and José Ramos Mejía were responsible for organizing the Sociedad de Antropología Jurídica (Legal Anthropology Society), which published an influential bulletin in which this ongoing discussion was encouraged. The journal *Criminología Moderna* (Modern Criminology) was an outlet for socialists and anarchists. The Archivos de Psiquiatría y Criminología (Archives of Psychiatry and Criminology) became the most important journal on this topic not only in Argentina, but throughout Latin America. Several monographs also reflected the new perspective on Darwinian ideas. Francisco de Veyga published his influencial *Estudios Medico-legales* in 1879, and Drago wrote *Los hombres de presa* in 1888. José Ingenieros's *La simulación de la locura*, published in 1903, would become the definitive example of how to apply Darwinian theory toward the understanding of society.[14]

It is probable, as Rosa del Olmo has affirmed, that Argentina was the first country to adopt the Italian criminological approach so wholeheartedly, and it was certainly "the first to put clinical criminology into practice in the penitentiary" (Olmo 1981, 21). Due to the already established interplay between science and government, the criminal justice system was motivated to implement the new theories, thus inserting the new evolutionary ideas into political practice. By the turn of the century Argentina would feel the influence of Darwinism at several levels, including in its legal and penitentiary systems, a development that provoked strong opposition by those who were alarmed by the degree to which materialism had become ascendant. Many of those who promoted science and remained avowed Darwinians were nonetheless concerned about the failure of new ideas to balance well with an older idea of civilization that was based on both material and spiritual principles.

By the end of the nineteenth century and the beginning of the twentieth, psychology began to attract the attention of younger scientists, leading to some highly original and interesting work (see chap. 6). In 1910 José Ingenieros published what is regarded as one of the first works on biological psychology based on Darwinian principles. His *Principios de psicología biológica* (Principles of Biological Psychology) attempts to do for psychology what Ameghino had done for phylogeny, to create a synthesis that served to organize a system of knowledge that was by now a characteristic of Argentine thought. In *Principios*, he weaved "a synthetic psychological system from positivist philosophy and physical chemistry, uniting mental phenomena at ontogenetic, evolutionary and social levels" (Triarhou and del Cerro 2006b, 1).

The Argentine Synthetic Imperative

Argentine intellectuals placed their hope in the universal culture of civilization because it gave them a sense of historical continuity, and an identity affirmed in shared values that masked the alleged imperfections of the Argentine population. But Darwin had changed the conceptions of time and history and thus changed basic constituents of the historical sense itself. Time was not universal, for different species evolve along different timescales. Each individual had its own history, its own contingent explanation. As Tony Bennett has explained, a Lamarckian sense of time "converted time's passage into a law of progress." But on a Darwinian understanding of history, "the mere lapse of time itself did nothing." Here time becomes genealogical, and "insignificant and anomalous details" become crucial elements in Darwin's "reconstructions of lines of descent and inheritance" (Bennett 2004, 50). Furthermore, the strange and particular were now more important than any commonality. Following Bennett again, we conclude that the study of history became a matter of tracing sequences of linear descent, completing and accumulating the data needed to understand evolution. The museum is precisely the institution to accomplish this purpose (50–51). But while in post-Darwinian Europe the function of this institution was to *gain* knowledge and complete the understanding of genealogical change, in Argentina it was to prevent the *loss* of the sense of self. By 1884 the new Museo de la Plata, directed by Francisco Moreno, had become a monument to evolutionary theory; "In keeping with Darwinian principles, Moreno set out to gather collections and design exhibits that would illustrate the entire course of evolution in Argentina, covering everything from fossil remains in local sediments to contemporary industry and arts" (Lopes and Podgorny 2000, 114).

But in understanding this world of loss, intellectuals felt the need for a philosophical system that sustained the prospect of continuous perfectabilty for the nation. It is for this reason that Herbert Spencer's work was of such importance. He lent Darwinian evolutionism a philosophical framework, in much the same way that Haeckel provided scientists a sense of philosophical continuity. Incorporating evolutionary theory, synthetic philosophy, and monism, the intellectuals analyzed in this book continued to operate within the tradition of totalizing narratives. In adopting such platforms, Argentine evolutionary thinkers were continuing a tradition in which natural and social worlds belonged to the same system. Spencer recognized the influence of Humboldt's *Cosmos* in the development of the idea of evolution, and he retained its central theses. His topics are "the universe, the earth, and life," and although he does not ignore struggle, "he certainly subordinates it to his own explanatory factor: Lamarckian inheritance of acquired characteristics, the concept of physiological division of labor, and that mysterious process which seems to be the key to all change—the transformation 'from homogeneity to heterogeneity'"

(R. M. Young 1985, 51). Even when Spencer had a place for struggle, "it basked in the light of Progress. This must be the real source of nature's energy" (51). According to this idea, the change from the homogenous to the heterogenous was unproblematic because, in the final analysis, progress was guided by a design, the recognition of which sustained hope in perfectibility.

In terms of the study of society, Spencer provided the needed totalizing system, one in which the march of progress toward future perfection emerged as an indisputable consequence of the scientific image of nature. Of course, the reality was that the whole miscellany of nineteenth-century European thought was not in the least coherent, nor was it designed with a view toward straightforward answers to the pressing problems of developing societies. In Argentina, as elsewhere in Latin America, the nation was in the process of formation, dealing with a culturally, racially, and ethnically diverse population. With regard to the needs of such a society, a theory constituted by analogies to variation and competition was not particularly helpful. Furthermore, new scientific discoveries created confusion among evolutionary thinkers. The importance that hard inheritance (as opposed to the Lamarckian, or "soft," inheritance of acquired traits) took on by the end of the nineteenth century and the division of evolutionists into neo-Lamarckian and neo-Darwinian camps constrained the possibilities of developing a theory that addressed both scientific and sociopolitical local needs[15] (see chap. 6).

As Keith Ansell Pearson explains, the work of leading neo-Darwinist August Weismann severed "Darwinism from its entanglement in Lamarckian dogma, so making way for the establishment of a strictly mechanistic and nonvitalist theory of evolution by placing all the emphasis on natural selection as a blind machine that guarantees the reproduction of human life from generation to generation in terms of unbroken descent" (Pearson 1999, 4). In Argentina the dispute around Weismann's ideas was related not only with scientific concerns, but also with the very principles that legitimized the existence of the nation. Because analogies stemming from hard inheritance appeared to emphasize a mechanical mode of evolutionary change, as opposed to the spiritual and vitalist ideas on which the nation had been founded, they left the human will impotent to direct the course of such change, which in turn made political will completely irrelevant.

The fight among neo-Darwinians and neo-Lamarckians also coincided with the threat of the United States's doctrine of Manifest Destiny and the creation in 1906 of the Monist League by Ernst Haeckel, both developments demonstrating the need to define national existence in essentialist and idealist terms so as to confront the mechanicism inherent in the renewed emphasis on natural selection.[16] This association between Darwinism and imperialism eventually led the most civilized nations to war in part because of the racialization of idealized national types. In 1914 Carlton Hayes, professor of political science at Columbia University, wrote an article about war and competition among nations. In it he asserts, "the popular misconception of Darwinism and the application of biological hypothesis to modern nationalism is lamentably not a peculiarity of the Germans." According to this author, Spencer and other English writers had had much to teach the United States about the "manifest destiny of the Anglo-Saxon race," the inherit superiority of "Anglo-Saxon" institutions and of the "Anglo-Saxon genius," the "white man's burden," "the struggle for existence," "and all the rest of it" (Hayes 1914, 706).

The rest of it includes the counterpart that emphasized the decline and degeneration of the Latin race, and its lack of evolutionary potential. Argentine intellectuals had already understood the destructive potential of these ideas. By the early twentieth century, it was clear to the scientists discussed in this book that evolutionism was marching in a direction that contradicted Argentine national interest. This perception lent further impetus to the search for an ideologically congenial philosophical complement to Darwinism.

In this vein, Angel Gallardo's (1867–1934) PhD dissertation in natural sciences, defended in 1902, expresses the author's disappointment with "the materialist conceptions that had dominated the science over the last century." In opposition, he proposed a more dynamic approach to the study of inheritance, based on the work of neo-vitalist botanist Johannes Reinke (1839–1931), which addressed "this transcendental philosophical problem" of the notions of matter and energy. He concludes that, after all, "we know matter only through the impression of our senses, that are due to manifestations of energy" (Gallardo 1908, 137). In citing Reinke, author of *Die Welt als Tat* (The World as Deed, published in English translation in 1899 as *The World as Reality*), Gallardo reveals his own interest in neo-vitalism, a school that would be influential in new philosophical approaches developed in opposition to neo-Darwinism. Henri Bergson, one of those who were influenced by Reinke's dualism, explained that by the beginning of the twentieth century the neo-vitalist movement was divided in two camps, "on the one hand [those who adhere to] the assertion that pure mechanism is insufficient," influenced by Hans Driesch (1867–1941) and Reinke; and, "on the other hand, [those who prefer] the hypothesis which this vitalism superposes on mechanism (the 'entelechies' of Driesch and the 'dominants' of Reinke, etc.)" (Bergson 1911, 42). As a member of the first of these camps, Gallardo's calls for the philosophical renewal of science mesh well with his call for the youth of Argentina to return to identifying science with Fatherland (*patria*), even in the midst of the present philosophical crisis. "Let us labor in the field of physical science ... as well as in applied science, which is of greater immediate utility and closer to the country's present needs. In return we may expect honor and profit for our Fatherland [*patria*], since scientific work is also patriotic work" (Gallardo 1980, 138).

Gallardo's dual interest in science and politics demanded a view of heredity that balanced internal and external conditions, against the exclusive emphasis on hard inheritance so prevalent among neo-Darwinians at the beginning of the twentieth century. As Sergio Cechetto notes, he preferred this approach to a general theory of inheritance, "since he found it more satisfactory to give a biological-moral character to the effort and ingenuity of individuals striving to overcome the grave and decadent legacy of their ancestors" (Cecchetto 2008, 133). Gallardo's association with some of the leading scientists of the day and his contact with Dutch botanist and pioneering geneticist Hugo de Vries (1848–1935) made him well aware of the political and social implications that an emphasis on strictly deterministic language could have on a society that had invested its future in the possibility of creating a new, civilized population. By 1908 Gallardo began to reflect on Mendelianism, the other important factor in the dismissal of soft inheritance in the explanation of social change. The prestigious French *Revue Scientifique* took notice of Gallardo's perspective on the relationship between Mendel's laws and the statistical distribution of traits in adult populations. "According to M. Gallardo, Mendel's law was discovered and proved by means of careful experiment, but it may well disagree with the statistics of adult populations that have undergone strong selection as a function of the differential mortality of their constituent groups.... [However,] the lack of agreement between theoretical predictions and statistical results for adult populations formed outside experimental conditions in no way refutes the law" (Delage 1908, 285–86).[17] The same paper cited in the *Revue Scientifique* was also remarked on the following year in *Nature*. "*Inheritance* in biology forms the subject of an address delivered by Mr. Angel Gallardo before the Instituto de Enseñanza General at Buenos Aires. Of this address, which has been published in the *Biblioteca* of that institution, we are indebted to the author for a copy" (*Nature* 1909, 105).

In this reference we find Gallardo addressing one of the problems that would exercise Sewall Wright, J. B. S. Haldane, Ronald Fisher, and other founders of population genetics in the 1920s. But in order to deal at the same time with patriotic and scientific needs, Gallardo, like most of his contemporaries, had to balance contradictory points of view. Argentine intellectuals felt a need to unify scientific insights in such a way as to affirm both nationhood and the coherence of civilized thought.

Florentino Ameghino, as we have noted, is another figure whose nuanced and evolving attitude toward Darwin merits careful scrutiny. By the turn of the nineteenth century, his philosophical ideas reveal a less materialistic perspective. For example, *Mi Credo* (My Credo), a talk delivered and published in 1906, includes a defense of eternal progress and a view of the universe as divided into four "infinites": matter, time, space, and movement. In trying to solve the various epistemological problems that resulted from Darwinism as understood in Argentina, he introduces a type of thinking that some associated with mysticism, seemingly contradicting some of his scientific principles. But, as we have noted, to call this attitude "Lamarckian" would do injustice to the complexity of the blend of ideas that it encompasses. In our view, the philosophical work in which Ameghino expresses a metaphysical interest divorced from his otherwise Darwinian perspective is motivated not by science, but by the broader cultural environment of the end of the nineteenth century. We agree with José Ingenieros that what characterizes Ameghino's philosophical views are his optimism and idea of perfectibility, both connected with the culture that dominated Argentina at this time.

> [Ameghino's] moral position was thoroughly optimistic. He let himself be swept away by his imagination in his predictions regarding the future human longevity, what he called "immortality" in metaphorical terms, more appropiate to poetry than to science.
>
> ... And, without leaving Nature, he imagined a God born from Nature herself: the perfected man of future humanity. (Ingenieros 1957, 179)

Ameghino's work as a paleontologist followed in Darwin's footsteps, as his interest in variation and extinction shows. But paleontology was also at the service of the spiritual continuity of Argentina through its defense of the country's foundational values. Ameghino supported this enterprise, and his philosophical ideas connected spiritual notions with the work of patriotic scientists who named, classified, and organized the evolutionary legacy of the country. "The few tens [of fossil mammals] recorded during the previous period were discovered and described by foreign naturalists, among whom the names of Owen and Darwin stand out. But I am pleased to announce a point of great honor for our country: over the past two decades, nearly all of the extinct mammal species to acquire their citizenship papers in this nation, a nation so hospitable to science, were discovered, catalogued, and described by Argentine explorers and naturalists" (Ameghino 1904, 516).

We are not the first to notice Ameghino's curious duality. David Rain Wallace also notes that, as "an enthusiastic Darwinian," he named "an unusual number of species, implying that a vast amount of evolution had occurred." His Darwinism "didn't convince him, however, that the mastodons, horses, and other 'great quadrupeds' in Argentina's Ice Age deposits were superior immigrants from the north. Indeed, his own collection's abundance suggested to him that North America's Ice Age mammals might just as well have immigrated from the South" (Wallace 2005, 82). Ameghino's emphasis on the evolutionary relevance of the region now known as Argentina derives in part from the importance of demonstrating Argentina's larger evolutionary significance. Instead of simply accepting the prevailing assumption that most American fauna had originated in the North, which would yield the privileged evolutionary position to the territory of the United States, Ameghino offered an alternative origins thesis of his own. By 1906 he would even try to prove that humanity's origins could be traced to Patagonian marsupials. "South America's ungulates, carnivores, primates, and other placental mammals had spread out to occupy the rest of the world, he concluded, after the sea had covered Patagonia at the end of the Cretaceous, some 65 million years ago" (84). This position gained him plenty of admirers, though not for the right scientific reasons. In 1919 Ingenieros himself criticized the "ridiculous collective vanity, disguised as patriotism" that led some to defend the paleontologist "because he attempted to situate the cradle of humanity on Argentine territory," an obvious scientific mistake (Ingenieros 1957, 153).

Ameghino's speculation shocked the established scientific community, because he was challenging not only the science of the time, but also the hierarchy that sought the origins of civilization near the most civilized and evolutionarily advanced populations. Ameghino was wrong, but we need to understand his mistake in its proper cultural context. We are not implying that Ameghino tried to force science to confirm political agendas, but rather that the scientific culture of the time left certain trajectories open. For, as Wallace recognizes, at the center of the dispute was not only scientific evidence, but also the imperialist presuppositions that consigned Latin America to a lower place on the evolutionary scale than those countries identified with progress. After all, "Argentine claims to mammalian priority disturbed North Americans even more than Europeans. Manifest Destiny was still in the air, and empire on the horizon" (Wallace 2005, 85).

It was precisely this association of imperialism and science that turned many intellectuals against the deterministic biological approach that had gained popularity by the end of the nineteenth century. In 1907 the criminalist lawyer Antonio Dellepiane (1864–1939) also reflected on the current state of evolutionary thinking, particularly as applied to the study of societies. According to Dellepiane, the action of the organicist hypothesis of the past had been "negative and disturbing," because it had "contributed to wasting souls of superior merit" who had relied so strongly on this conception of society that they had abandoned reasoned argument entirely.

> [They] believed they might understand social phenomena by describing them in biological terms, or they pretended to explain them through the unilateral and exclusive action of some biological law, such as the law of adaptation or that of sexual selection, without seeing that a deep chasm separates the physiological process of vital adaptation from the obviously psychological process of the adaptation of individuals to the social environment, and without noticing that sexual selection in the human species is something very different, since it is conditioned by the most diverse motives—affective, economic, political, religious, etc.—from the phenomena of the same name in species inferior to ours. (Dellepiane 1907, 67)

As Dellepiane makes clear, by 1907 there was widespread recognition of the inherent limits of the biological understanding of societies. By then Spencerian analogies that relied on "soft" inheritance, the inheritance of acquired traits, had begun to be regarded with suspicion. The development of psychology and the understanding of human adaptation as a psychological process proclaimed a new way of addressing the specificity of human nature and the importance of the human mind in the process of evolution. In fact, the philosophical failure of Spencer's synthetic system triggered a renewed interest in psychology and philosophy, particularly German philosophy.

Further demonstrating that the eclecticism of Argentine thinkers was not a matter of mere dilettantism, Dellepiane explains that, faced with a choice between the salt of evolutionism and the yeast of "German metaphysical schools and their admirers in different countries," the Argentine academy "chose both," reflecting the "spectacle, interesting to few, of a perplexed philosophical consciousness fighting to reconcile antinomies." In the course of the resulting struggle, Kant became an important figure for Dellepiane and his contemporaries. "At the bottom, the solution to the problem was the resolution of this dynamic antinomy: the determinism of natural phenomena is not absolute and free causes exist; there is no freedom, and everything happens according to natural laws. This antinomy, according to Kant, has its origins in the possibility of the dual use of the understanding, the empirical and the transcendental, of which only the former is legitimate" (Dellepiane 1907, 7–8).

The return to this philosophical source permitted a restoration of the human will in the process of historical development. Teleology is asserted, not in a religious sense, but in the human determination to conquer nature and make it obey. The pre-Darwinian idea of nature as subservient to human desires in the process of civilization once again becomes relevant. Dellepiane's interest in finding a middle ground between metaphysical and materialistic concerns finds another example in an address given by Carlos Octavio Bunge to law students in 1907. In it he admits that the "transformist theory" had suggested a "new concept of man, of psychology, and of society. This concept changes in a radical way the old points of view, undermining the old axioms and prejudices of eighteenth-century philosophy" (C. Bunge 1907, 155). But Bunge continues to admire the precision of some of these philosophers, asserting that Kant and Hegel would have been positivists if they had learned of post-Darwinian ideas. He praises "the degree of exactitude that, with their sometimes extravagant speculative methods, some metaphysical thinkers achieved. In a certain famous passage, Kant anticipates Darwin's theory; his conception of law and justice coincides in its fundamental points with Spencer's modern conception" (155). It was thus a mistake to discard metaphysical approaches in favor of purely positivist conceptions, where by the latter he also means post-Darwinian materialism. According to Bunge, "none of these tendencies has an exclusive monopoly on the relative truth accessible to our limited senses and to our narrow intelligence as men" (155). He concludes by affirming that it was possible to conceive the truth through "opposing and varied means, even highly idealized, speculative form, though this is more difficult every day because of the enormous volume of contemporary scientific information" (155–56).

Even sociologists who had been opposed to any role for metaphysical speculation began to recognize that a new synthesis was needed in light of the philosophical debacle that had arisen in the wake of the neo-Darwinian resurgence of natural selection, with its concurrent demotion of Herbert Spencer's work. In 1911 the lawyer and historian Ricardo Levene (1885–1959) offered an overview of the ideas that had dominated the intellectual life of Argentina since the mid-nineteenth century. He explains that the influence of Spencer, particularly among sociologists, could not be separated from that of Darwin's theory. According to him, "Spencer transplanted to the camp of Sociology the theory that Darwin had developed for specifically biological phenomena." In this way, the theory of evolution "was the marrow of the Spencerian doctrine, that organic evolution is no more than a particular case of general evolution" (Levene 1911, 238). Inquiring into the specific contributions of "Darwinian selection," Levene asserts that it had sanctioned the reconstruction of "the genealogical tree of species, demonstrating evolution and delivering a formidable blow to supposed spontaneous generation." The most important concepts introduced by this doctrine were "the struggle for existence, the survival of the strongest, the disappearance of the weak, variation, natural selection, and the origin of species through [the preservation of] useful variation" (238). Darwin had "generated passion among all the chosen spirits of the era, detractors as well as supporters." The central question was "Might the struggle for life be the cause of the conservation of those animals that possessed the best qualities, suppressing the weak?" (239).

The philosophical problems that arose out of the division of evolutionary theorists into neo-Lamarckian and neo-Darwinian camps demanded, in Levene's view, a reassessment of particular scientific claims. He sustains his faith in science and anticipates that in the future an "eclectic scientism will bring harmony to this disagreement on details, and the sectarian divisions between Lamarckianism, Darwinism, neo-Lamarckianism and neo-Darwinism, will disappear to give way to a broader generalizing doctrine" (239).

This desire for a synthetic principle capable of uniting disparate philosophical lineages would continue in the years to come. In 1914 José Ingenieros reaffirms Darwin's status as the most important figure in the nation's intellectual history, followed by those who applied his insights to specific fields: Claude Bernard and Rudolf Virchow in the medical and biological sciences; Auguste Comte and Achille Loria in the social sciences; Herbert Spencer and Théodule-Armand Ribot in philosophy. These names were often followed by those of "[Jean-Baptiste] Lamarck and [Ernst] Haeckel; [Jean-Martin] Charcot and [Louis] Pasteur, [Émile] Durkheim and [Gabriel] Tarde, [Friedrich] Froebel and [Johann Heinrich] Pestalozzi, [Wilhelm] Wundt and [William] James." In the work of historians, "the constant influence of [Hippolyte] Taine and [Ernest] Renan" was evident (Ingenieros 1914, 89).

In Ingenieros's view, this unique blend of sources indicated that Argentine thought was preparing "to ascend to a philosophy that places science on a firm foundation, and derives from it the ideals of a nascent race." In his view, the twentieth century would give rise to "a system of scientific philosophy that may contain the 'new meaning,' unique to Argentina, of the future culture" (89). In this passage it is clear that what Ingenieros saw as the future culture of the country was related to the creation of a definitive race under the control of a science simultaneously based on ideals and experience. Darwinism needed to be adapted to local realities, which required a philosophy that contemplated these special needs. Against the old, rigid, aprioristic idealism, "the scientific philosophy will oppose an experimental idealism in which the 'ideals' are constantly renewed, plastic, and as evolving as life itself" (90). Finally, from the philosophical point of view, "argentineity" (*argentinidad*) would consist "in the new meanings that the emerging race of this part of the world can impress on human experience and ideals" (91).

The result of this process was a blend of ideas of which Edward Schaub, a professor of philosophy at Northwestern University, took notice in 1928, asserting that in Argentina "European thought acquired a vernacular coloring which is in itself of no small interest to the historian." According to him, in this country "there are elements of thought that are specifically Argentinian, despite their European roots" (Schaub 1928, 603). Schaub clearly identified the different lineages of the ideas that were important in the country and correctly noted their peculiarities as the result not of insufficient attention to European sources, but of attempts to build a system of thought that would work in the specific context of Argentina.

Schaub describes the different influences as deriving, in chronological order, from the Scholastic philosophers, French Enlightenment thought, European romanticism, and finally positivism. But he recognizes that uses of the latter "emphasized chiefly the negative aspect of the thought of Comte and Spencer." He uses the word *negative* because, in assimilating this thought, Argentine thinkers have chosen to neglect the very logic of positivism, particularly the more dogmatic among those who had devoted themselves "to the metaphysical interpretation of Darwinian biology in the manner of Haeckel." Schaub also mentions Ameghino and his "bold profession of philosophic faith on the basis of a mystically materialistic cosmology" (603).

The article concludes with the strong assertion that "Haeckel is one of those chiefly responsible for the vague philosophic thought of Spanish America. A noisy and decrepit Haeckelian senility found in its ranks educational scientists and literary psychiatrists" responsible for a proliferation of pamphlets on the "future of philosophy" that would doubtless prove "valuable documents for the future historian of the philosophic dilettantism on the coast of the Plata" (604). But what Schaub missed was that the adoption of different, often contradictory, currents originated not in Argentina, but in the very centers of scientific thought.

The thinkers we will analyze in the second part of this book were not wrong in pointing out the need for a scientific synthesis, because the trajectory of evolutionary theory itself ultimately led in that direction. But in gesturing in this direction, they also reflected the cultural contingencies of a theory built on analogies. They related the process of synthesis to nation building, adapting science to the present needs of their country, as did their counterparts in Europe and the United States. But while in the centers of scientific research cultural interventions were disguised within the main currents of the evolving scientific process itself, Argentine thinkers were not able to claim the same centrality on behalf of their own scientific work. As such, their attempt must inevitably appear to Schaub as peripheral and insignificant. The eclecticism of Argentine thought was, in his view, the product of mere misunderstanding and not, as we have been arguing, of deliberate correction.

Carefully analyzed, the principles introduced in *Origin of Species* in 1859 could be seen to subvert traditional notions of evolution and progress. The subversive nature of Darwinism with regard to the old Enlightenment ideas was also felt in Europe, where thinkers like Huxley ended up attempting more spiritual, less materialistic interpretations, because they were also bothered by the implications of evolutionism. But Huxley was seldom branded a dilettante, as Argentine scientists would be. Nor was Huxley laboring under the obligation to weave his own work, together with that of his European scientific peers, into a coherent whole. The chaos and contradictions brought in by the publication of *Origin* meant that the faith in the unity of science and the coherence of civilized thought were not as real as the defenders of European ideas in Argentina had come to believe. Thus, in order to save the foundation of the future civilized nation, they had to create a truly civilized universalism that erased the different lineages of European thought. In forging such a creative synthesis, they operated under what we call the "synthetic imperative." In their attempt to make Darwinism fit with ideas viewed as essential to the dominant ideology of any modern country, Argentine Darwinians created the unique synthetic blend Schaub captured. Bridging the French Enlightenment tradition that preceded Darwin with the new evolutionary ideas tainted with German Idealism, they attempted to maintain the existence of a biological design that supported their faith in the country's future. And so we find either dark positivism or optimistic Darwinian evolutionism—either way, always sustaining the future of the nation.

PART II

CHAPTER FOUR

The Culture of Extinction

In evolution by natural selection, every taxon, in every generation, confronts the exclusive disjunction between *evolution* and *extinction*. Within the confines of Darwinian evolutionary theory, the two are nearly logical complements. In the constitutive analogies of Darwinism, this duality plays a central role. In this chapter, we explore the fate of those analogies in a social milieu that represented itself not as blessed with evolutionary promise, but as threatened by extinction.

At the time of his death in 1984, the famous paleontologist and evolutionary theorist George Gaylord Simpson was at work on an article on *extinction*, a key scientific term that had been redefined in the context of the Darwinian revolution. In this article, he explains that before Darwin we can find descriptions of evolution without extinction (Lamarck) or extinction without evolution (Cuvier), but only after Darwin can we see the concepts of evolution and extinction working together.

Cuvier was known to many of those young intellectuals who, by the 1830s, had joined to fight Rosas in support of Enlightenment ideals. As we have discussed, he was influential in the work of Francisco Muñiz and Juan María Gutierrez. In Gutierrez's 1835–1836 article on the *Megatherium*, he discusses the French naturalist's ideas. They provide the context for his own analysis of change in nature, broadly catastrophist in outline (Gutiérrez 1835–1836; repr. in B. Martin 2005–2006, 2:9). "There is a branch [of the natural sciences] in which it is not sufficient to study the objects of which nature is now composed. One must also consider those that once existed but have disappeared forever, in one of those general perturbations that afflict our globe, or in some partial cataclysm, as recorded by tradition or preserved within the successive layers of the earth like so many pages in the history of its formation" (B. Martin 2005–2006, 2:10). In Simpson's words, Cuvier's paleontological evidence had demonstrated that "the history of the earth and of its animals and plants was punctuated by a series of revolutions, in each of which many extinctions occurred." In this sense there was "a whole sequence of mass extinctions followed by quite different genera and species" (Simpson 1985, 410). By contrast with Cuvierian catastrophism (extinction without evolution), Lamarckianism (evolution without extinction) "is based on the assumption that changes of structure produced by the activity of the adult organism can be reflected in the material of heredity and passed on to the next generation" (Bowler 2003, 243).

As we have noted, Lamarckian ideas were important for the Argentine intellectuals who placed such stock in evolution as a way to overcome primitivism. They were particularly attracted to the possibility that culturally acquired traits could be transmitted to future generations. As Noël Salomon has explained, "both Sarmiento and the majority of his contemporaries among Argentine intellectuals were susceptible to the reasoning of eighteenth-century French philosophers, and to that of the French historians prior to 1850" (Salomon 1984, 5). In short, many of Lamarck's own sources of inspiration were important to Sarmiento and his generation. They, in turn, invoked these sources in their attempts to anticipate social change. When they encountered new ideas in tension with older influences, they engaged in creative, holistic attempts to reimagine, and unify, the whole complex of thought reaching them from the civilized world.

By 1850 the members of the Generation of 1837 mostly believed, very much in a Lamarckian way, in a ladder or tree of life that forked to form terminal branches. With Lamarck, they "did not consider this as extinction. For him the series, ladder, or chain that led to birds" still lived, "*as* birds" (Simpson 1985, 410). The French naturalist went no further in his study of the "lost species" because he never clearly "conceived [them] as ancestors or related species that could explain the historical process of evolution." Species "were always changing but never ceased to exist" according to this view. Lamarck did not need a notion of extinction, because his thought remained impregnated with Deist notions. "The idea of the linear tendency in complexity ('echelle,' 'chaine graduée') is influenced by the notion of the 'Chain of Being,' which excluded the possibility of interruption of communication in nature" (Tassy 1981, 198).

Darwin disrupted this idea with his emphasis on extinction, the interruption of lineages. In doing so, he introduced terrible doubts among those intellectuals who wanted to organize the modern nation in the hope that its evolution might prove a generative process of continuous progress. The Darwinian introduction of selection in place of assimilation transformed the very idea of civilization. In blending inheritance there was no selection, but constant fusion, and in the final result all the elements were present, though they were not all visible.[1] But with selection something was lost, and this waste implied that evolution was not a fully harmonious process, but a painful one in which both life and death were present. Leading intellectual Ernesto Quesada (previously discussed in our introduction) described this novelty in the new evolutionary thought as follows.

> This simple fact is what Darwin has called natural selection, the survival of the best suited. It is also a fact, equally easily provable by anyone, that not all the fruits of one and the same trunk are identical, nor do they find themselves in identical conditions. It is precisely these variations among offspring that allow for the selection of the better, and the destruction of the worse, such as we regularly practice among the plants in our gardens and the domestic animals in our fields. Such choice, exercised by us in these examples, is exercised by nature among wild species, in such a manner as to allow only the best exemplars to survive, those best suited to the environment in which they develop. (E. Quesada, 1907, 12–13)

As Quesada notes, the logic of the Darwinian theory is grounded in the dual processes of evolution and extinction, both explained by the mechanism of natural selection. The process of extinction had been recognized, before Darwin, by Cuvier—but "it was Darwin who among many other things gave us evolution *with* extinction" (Simpson 1985, 411; emphasis in original). What's more, the complementarity of extinction and evolution meant that extinction, when observed in the historical present, could be pointed to as evidence for evolution in action.

Darwinism *naturalized* extinction. For Cuvier, extinctions were typically associated with catastrophes, departures from the ordinary natural course of events. For a pre-Darwinian naturalist like Francisco Muñiz, the prospect of extinction without a geological catastrophe could arise, but only as a consequence of the operation of human forces unchecked by nature. For example, in the study of the ñandú discussed in chapter 1, Muñiz expresses his concern for the future of the species, endangered by unbridled hunting on the Pampas. On more advanced estancias the ñandú was less endangered, as it enjoyed "protection and security" (Sarmiento 1885b, 179). The extinction of the ñandú as a result of human predation would not be a *natural* event. And in protecting "this noble and valuable American species" (179) against abuses, landowners were helping to *enforce* the natural order, in opposition to the ignorant barbarians who abused it.

Darwinism inverted this reasoning. Because it allowed, in general, for extinction to be read as evidence for evolution, it sanctioned a very different interpretation of the loss of Argentina's indigenous species—and of its indigenous human populations. Now such events could be rationalized as an unavoidable consequence of the country's modernization; the disappearance of those who could not keep up with this process was proof of progress. Such progress was what the Enlightenment-schooled generation of Argentine liberals who defeated Rosas so yearned for. But when, in the wake of Darwin, progress was measured by extinction, its meaning was subtly changed, because it could no longer be understood as motion toward final unity, the harmonious blending of diverse ethnic types.

To be sure, in Argentina as elsewhere in Spanish America, the idea that a human population might be wholly exterminated was hardly a new one. In his Jamaica Letter, Bolívar had called the Spaniards "exterminators," their victims the native peoples annihilated during conquest and colonization. Here, too, Darwinism imposed a meaning shift. Earlier in the nineteenth century, civilization had been understood as a social force operating through assimilation and cooperation, transforming the *abstract* body of a nation, its culture and ideology. By the end of the century, the progress of civilization had come to require the replacement of the individual *physical* bodies that comprised the nation. Its evolution must now be conceived, in Darwinian terms, as entailing *waste*.

The diversity of the Latin American population had been seen as a problem since the early days of independence, but the typical solutions to it involved the blending of all individuals into a single national family. As Doris Sommer observes, "Even the most elite and racist of the founding fathers understood that their project of national consolidation under a civil government needed racial hybridization" (Sommer 1991, 123). The new science undermined the feasibility of such union. Darwinian population science appeared to indicate that populations needed to be reproductively isolated, so as to allow the most favorable traits of each to be identified and selected. Their departure from post-Independence attitudes could not be more dramatic. For Sarmiento and Alberdi, isolation had been the *source* of barbarism. As Gabriela Nouzielles has observed, the following generation underwent a shift from liberal to ethnic nationalism, a transformation we explain, in turn, by the change in the meaning of civilization wrought by the Darwinian revolution (2000, 18). Nouzielles's observation on the state of *porteño* culture at the end of the nineteenth century is of a piece with the post-Darwinian anxiety of extinction. Concern with the present and future consequences of miscegenation during the Colonial era, shared by members of all the educated classes, masked a deeper preoccupation with the prospect of an end to reproductive continuity, the price for poor mate choice on the part of their ancestors. The male fantasy of perpetual fecundity identified by Doris Sommer in her analysis of the founding narratives of the nation had been shattered by the Darwinian identification of the progress of civilization with biological evolution and its inevitable complement, extinction.

The authors discussed in this chapter all shared a fear for the continuity of their posterity. The constitution of 1853 had declared the openness of the population essential to the destiny of the new nation, calling for massive European immigration. The certainty with which the elites embraced this policy began to waver slightly after the publication of *Descent of Man* in 1871, in which Darwin appeared to open the door to various forms of soft inheritance related to sexual selection, but the possibility—indeed, the inevitability—of extinction as a by-product of progress, remained entrenched. In his "Vision y realidad" (Vision and Reality), Florentino Ameghino showed clear appreciation for the darker side of population replacement, inverting the triumphant Tree of Life image, as deployed by Haeckel and other European scientists, turning it into a Tree of Death that litters the ground with its cast-off branches, each representing an extinct group.

The prospect of extinction also appeared to undermine the altruism that Alberdi and his generation had called for in their drive to repopulate the country with European immigrants. For Alberdi and Sarmiento, public-spirited altruism had been a necessary condition for the construction of a civilized nation, a process that, after all, required that the country be given, without recompense, into the secure hands (and reproductive organs) of superior European males. While such altruism had been conceivable under the moral teleology of an earlier era, in Darwinian terms, it had to be understood as costly, "apparently too costly to have been favored by natural selection," and thus a poor foundation on which to rest public policy (Cronin 1993, 348).

Extinction in Action: The Disappearance of the Indians and the Afro-Argentines

The broader cultural importance that natural selection and its consequence, extinction, had acquired in late nineteenth-century Argentina is nowhere more evident than in frequent reflections on the "disappearance" of both the Indians and the Afro-Argentines. Along with such observations, we see growing efforts toward preservation—not of the threatened peoples themselves, but of their remains, to be displayed in museums as evidence of natural selection in action. Whatever the significance of museum culture in Europe and North America, in Latin America it was closely tied to the preoccupation with extinction on the part of national elites, who saw themselves, at least potentially, as next in line. The disappearance of undesirable populations was worth paying attention to as a happy sign that nature was working in the service of national progress, but at the same time, such attention could not help but raise uncertainty as to who, among those remaining, would stay to contribute to the nation of the future. Darwinism disrupted the traditional founding narrative of the generational continuity of the family; on a Darwinian genealogy, some lineages might persist, while the continuity of others was interrupted. The problem lay in predicting which were destined to persist, which to cease.

As we have discussed, post-Darwinian logic treated extinction as the counterpart of evolutionary progress. Under Cuvierian catastrophism, the extinction of a species might occur, but only as part of some terrible cataclysm, a disruption in the ordinary, harmonious course of nature. Human activity might produce such disruptions, but only in its most barbaric excesses; civilization, by contrast, represented harmony and life. But after Darwin, civilization, like other evolutionary progress, came at a cost. Where such costs had been recognized before, especially with regard to the destruction of indigenous populations, they had always been mitigated by the representation of civilization as an assimilating force. In the blending of skin tones, the white would overwhelm and absorb the dark. Now, however, the possibility of atavism, as recognized by Darwin, meant that so long as dark-skinned individuals had contributed to posterity, their traits might resurface in subsequent generations. The only way to ensure evolutionary process was for the bearers of such traits to be eliminated, not assimilated.

Prior to *Origin*, the elimination of the uncivilized could be understood as demanding simply the greater dissemination of civilized ideals. For those who adopted these ideals, assimilating to the culture of civilization, hope remained. As for those who succumbed to contact with civilization before embracing it, they deserved every sympathy. Consider, for example, Frédéric Lacroix's 1841 description of the Indian response to European diseases, such as smallpox. "The fear of contagion frequently leads the Patagones, like other southern nations, to become more inhumane. But who would not forgive them for reacting thus, they who have seen half their people laid low by smallpox in consequence of their contact with Whites? They see this disease, brought from Europe, as a special effect of a malign spirit, passed successively from one body to the next" (27). In this account, the European is the bearer of disease, and the Indian, conscious of the dangers of contact with whites but unable to prevent it, evokes compassion. But natural selection turned the replacement of the uncivilized by the civilized into a biological process, as exceptionless as any law of nature. There was nothing those destined for culling could do about it and no rational reason for anyone else to sympathize with them.

Compare Lacroix's view with the clinical detachment of Domingo Sarmiento in a November 1879 article on the Indian population published in the newspaper *El Nacional*. By this time, as we have noted, the campaign of extermination against the Indians of the Pampas and Patagonia was in full swing. Commenting, in post-Darwinian terms, on the losses they had suffered, Sarmiento observes, "The savages will be unable to recover for many years to come; and it is a fatal law of savage life, that they can never recoup their losses, for once they have come into contact with civilized peoples, they are condemned to final extinction. Each encampment represents a tribe about to disappear from the face of the Earth" (Sarmiento 1900h, 284). In another article in *El Nacional*, Sarmiento challenges the assertion by French correspondent Alfred Ebelot, who had fiercely criticized the campaign against the Indians, that the latter showed great resistance to "fatigue and provocation." In response, Sarmiento insists that, "among the tribes, all those whose resistance is anything short of extraordinary die young." This, he claims, "is the struggle for existence in full force. Nor is their ability to live in the open air, spend weeks on horseback, sleep naked on the ground, or survive the heat and the cold any proof of great muscular strength. Theirs is mere negative vigor; they are hard, but not muscular, and any sort of work quickly defeats them" (331). Where Ebelot charges the Argentines with atrocities, Sarmiento reminds his readers that the war currently underway is a *natural* conflict and the subsequent disappearance of the indigenous population its *natural* consequence. Within the limits set by nature, the behavior of the Argentine forces is even humane; for example, "The Indian children distributed among [white] families are happy in their own spheres, and frequently become the household favorites. The affections of their new families quickly replace those whom, in the course of nature, they so quickly forget" (331).

It is worth noting that later accounts of the Desert Campaign never refer to the massacre or displacement of the indigenous population, but only to its "disappearance." An extinction event, *qua* consequence of a natural law, is not the sort of thing for which one can attribute responsibility or, indeed, agency; no one is to blame. The indigenous and black populations simply cease being visible, and their invisibility is explained by reference to the process of natural selection. In an October 15,1885 debate in the Chamber of Deputies, Representative Dávila argued for reducing troop levels along the Chilean border, on the grounds that "now that the Indians have disappeared, we must question the necessity of maintaining a cordon of soldiers on such a peaceful frontier" (Argentina 1886, 502). A few scant years earlier, the mission of these same soldiers would have been clear; now, it is not that the mission has been accomplished, but rather that its urgency has abated of its own accord.

Of crucial importance in this context is the discourse surrounding the project of creating the future Argentine race. We see this discourse as grounded in two constitutive analogies drawn from post-Darwinian thinking: the analogy of extinction and the analogy of sexual selection linked to the return to idealist types through Haeckel's interpretation of Darwin. Whatever Darwin might have thought, in Argentine racial discourse, natural selection was viewed as a matter not merely of survival, but of improvement. Extinction, the inevitable counterpart to natural selection, could be read as evidence of evolutionary progress. When its victims were those unable to adapt to modern life, this was evidence of progress in the constitution of the new race. Such progress was a consequence of an implacable natural law operating in the selection of the best individuals. There are, of course, intimations of such teleology in Darwin, who, in what Stephen Jay Gould has called his "greatest failure of resolution," could never entirely shake the Victorian faith in progress (Gould 2002). These were greatly amplified in Argentina, where Darwinian ideas were self-consciously and often forcibly reconciled with Enlightenment and Romantic ideals predating Darwin. In the resulting synthesis, species tended toward perfection. Imperfections, and the organisms bearing them, would disappear along the road to the concrete realization of ideal types. The logic of this approach permeates politics—especially the racial discourse to which we have alluded—science, and popular culture.

In all three spheres, we find the same emphasis on disappearance—whether of individuals, groups, objects, or processes—as a function of the logic by which such loss is a necessary complement to evolutionary gain. In the political and cultural spheres, it originates in the popularization of Darwinism by sociologists. In the more narrowly scientific sphere, it is expressed by the attention devoted to documenting and classifying species—or human nations—as extant, extinct, or vanishing. In paleontology, Ameghino led the way in such efforts; in anthropology, Moreno.

Besides the beleaguered Indian nations, in this period two other groups were widely perceived as well on their way to disappearing: the *criollos* (creoles) and the Afro-Argentines. They became the subject of an extensive literature of racial nostalgia, for which the evolutionary synthesis described above served as background. One contributor to this genre was Vicente Quesada (1830–1903, the father of Ernesto), who penned the "Mi Tío Blas" (Uncle Blas) stories under the pseudonym Victor Gálvez in 1882. Himself a member of the doomed *criollo* group, he expects, or hopes, that disappearance will be mediated by assimilation, a long-term outcome of blending inheritance against the harshness of natural selection. The Tío Blas narratives are replete with tension between the promise of modernization and the sense of loss evoked by its success. Their protagonist is a shopkeeper, "mannered and cultured, who waited upon the ladies like a complete gentleman." Blas Tixera is "dark complexioned, wore a frock coat, and never worried about appearances." He forever laments that his "were the good old days, while today was the time of go-betweens, of gossip and ostentation [and] perpetual carnival" (V. Quesada 1882, 225–26). "And what do these bring you?—My good uncle does not understand that this is progress. He says that the boundaries imposed by good breeding, the noble gallantry of the past, the respectful courtesy with which ladies were always treated by the best people, those known as *decent folk*, have all been shattered" (228; emphasis in original). Don Blas belongs to "one of those types that was disappearing" with the transformation of Buenos Aires society (229). In him, Quesada presents a nostalgic vision of a group whose fate was already sealed, a group to which he himself belonged. Blas misses the serious days of "the old country," in which "respect for authority entered through the eyes." The powerful were "not like common men, but had authority dressed in the external signs" of their offices (231). The importance of visible symbols is apparent. In the old days, one would never have seen "magistrates with short trousers, black silk stockings, gold-buckled shoes, jacket and sword." Now, though, such men were everywhere in evidence, flaunting their sex, rather than commanding respect. In the past, such respect went without question. Authority was obeyed and gender-appropriate behavior strictly enforced. One story cites an 1809 regulation, signed by Viceroy Cisneros, imposing fines for lewd conduct, so as to assure that bathers on the Rio de la Plata's public beaches act "without exceeding the established limits for men and women" (234–35).

In the confusion of the present time, the old hierarchies blurred or erased, such limits no longer exist. Before, Blas "knew everyone by his name, and I would greet them as I passed: Good night, Don Pepito! May God keep you, Señora Andrea! Truly, we all knew each other, then. . . . There were no blasted books in any foreign languages; everything was for the *criollos*, in the language of our parents, with none of your mixtures or odd confusions. A spade was a spade, and nothing was frenchified." It is hard to avoid attributing Don Blas's annoyance to Quesada himself, along with the recognition that the very ideas he had himself championed and helped to establish as the national ideology were now destroying his world. "Everywhere I see Spanish, Italian, and French companies, but the *criollo* companies have disappeared from the scene. This is what I deplore" (236–38).

The Afro-Argentines were among the biggest losers in the racial redesigning of the national population. Curiously, for authors like Quesada, who remembered them as part of their vanishing world, the Black Argentines, once objects of derision, were now also the subject of racial nostalgia. The "same race whose savage songs and barbarous music caused so repugnant an impression [had] formed the ranks" of the armies that won independence (V. Quesada 1883a, 256). Theirs was surely "a type superior to the deformed black race of Dakar," though there was no obvious reason why it should be the case that "the black slaves in Buenos Aires were, physiologically speaking, superior to those savages of Africa who go about nearly naked. . . . The fact is that as the shape of their crania changed, their features grew more regularized, and though they retained the color, the flattened nose, the large mouth and kinky hair, the general form of their bodies improved so greatly that they became very attractive and well-proportioned black men and women" (258).[2] In the absence of any other explanation, the happy evolution of the African race in Buenos Aires could be taken as a reflection of the favorable conditions they had found there. As Quesada has it, they had prospered following emancipation, acquiring their own newspapers and other institutions, until "the colored race, as they call it among themselves, aspires to a level of culture that brings them closer to the white race" and deserving, if not admiration, then at least a certain paternal fondness. Quesada treats all those destined to disappear with similar affection,[3] including Don Blas, a member of the vanishing class of former slaveowners. As he depicts them, the Africans had "danced singing in their African dialects, as if in premonition that they were destined to blend with the white race, disappearing as a separate community within our capital" (V. Quesada 1883a, 257). Eventually, the African elements of such performances attenuated until they "were no longer imitations and recollections of African dances and songs, but adaptations of the religion and customs of the white race, with which they will ultimately be blended, by an inescapable law dictating that inferior races are assimilated into and absorbed by superior ones" (260). Indians and Africans had no future, nor did the author's own *criollo* constituency: "But the city of Buenos Aires holds in its bosom all these diverse elements, and against their current, it is difficult to sense the influence of the *criollo* race, the native element, which suffers the strangest transformations because there is nothing to preserve it against the stream of foreigners. What *can* be preserved? Language—and it is in this that the country's spirit may be saved" (V. Quesada 1883b, 247). The disappearance of the Indians was typically described, in stark Darwinian terms, as a matter of extinction. But the concept of blending inheritance thus continued to operate during this period. As we mentioned in our introduction, the lack of a clear explanation of the mechanism of heredity in Darwin's theory left room for different and contradictory ideas. And so Quesada and others reserved the somewhat gentler fate of assimilation for the Afro-Argentines and *criollos*, while progress was extinguishing the Indians.

This kinder mode of disappearance was in keeping with the general tone of literary nostalgia to which members of Quesada's aging generation were inclined in the 1880s. Other examples include José Antonio Wilde, who published *Buenos Aires desde 70 años atrás* (Buenos Aires 70 Years Ago), a memoir of the vanished city and its inhabitants, in 1881. A year later, Lucio V. López would begin to publish *La gran aldea* (The Great Village), a serialized novel dealing with the transformation of Buenos Aires society. In 1884, Miguel Cané published *Juvenilia*, an account of the childhood and adolescence of his contemporaries. By 1891 even the old Federalist and Rosas supporter Santiago Calzadilla had gotten in on the act, releasing his memoirs, *Las beldades de mi tiempo* (The Splendors of My Time). Where the previous generation of native-born Argentine intellectuals had sought to actively erase the barbarous colonial past, its heirs, the future of their kind uncertain, sought to reconnect with it. As exemplified by the liberal Vicente Quesada and his alter-ego, the somber preservationist Victor Gálvez, the identity crisis provoked by this prospect was not easily resolved.

José Antonio Wilde (1813–1885), a physician as well as a writer and public servant, was born in Buenos Aires, son of an English father and a French mother. As Wilde describes it in 1881, Buenos Aires has undergone vast changes during his lifetime. If a native son were to return now, after an absence of decades, he would be "astonished to see our old Black porters replaced by Basques, Italians, and Galicians" (Wilde 1961, 3). Their disappearance, along with that of other elements of the familiar past, is a deeply personal loss. A whole chapter is devoted to the world of the slaves, who were treated, so Wilde claims, "with real fondness" and admired for providing "brave, suffering, obedient soldiers of the first order," some of the "nation's finest troops" (38). But now "the Blacks are relatively scarce. Here and there one sees some veteran representative of this vanishing race, a monument eroded by time" (43).

Forgotten, it seems, are the derision and fear once expressed by members of Wilde's and Quesada's class, for whom the Blacks had been associated with the Rosist tyranny. Now their carnivals and songs are sorely missed, as are the cakes they once sold, a delicacy that has "been lost, like so many of our things." Ladies of "decent class" would actually lean over to buy these treats; now, Wilde asks, "What lady would do such a thing in this supremely aristocratic town? . . . None!" (43). Gone too are the Black laundresses, whom the author remembers "singing happily, each according to the customs of her nation . . . they seemed happy; they were never silent, and after certain remarks, doubtless very funny, a pealing laugh would resound; the laughter of the laundresses was very particular" (48–49). In those idyllic days, "[white] families would sometimes go out on summer afternoons to sit on the grass, drink *mate*, and enjoy the jokes and gossip of the laundresses." By contrast, observing today's laundresses, "the image is monotonous; the immense phalanx now occupying the place left by a race we saw slip away before our eyes like figures in a magic lantern, remains taciturn in its labors" (49). The population of African descent was now praised as "the first laborers, those who laid the cornerstones. If our predecessors were to come back to life, there would be much to amaze them—and also much to make them blush" (78). Following the Darwinian logic, progress could only come with its shameful counterpart, the elimination that resulted from the process of selection.

The Triumph of Darwinian Logic

By the end of the nineteenth century, the logic of natural selection, by which extinction represents the ultimate collective failure in the struggle for existence, had permeated nearly every aspect of Argentine culture. It is difficult to find an intellectual of any stripe on whom Darwin's theory had not left a significant mark, even among the conservative Catholic opposition. The extent of this impact was clearly in evidence at the first Assembly of Argentine Catholics in 1884, where Darwin's name featured prominently in several speeches. One of the principal goals of the convention was to organize opposition and articulate a response to the materialist currents that had come to dominate political and social life. For Catholics, Darwin served as an exemplar of an approach to ethics, and to altruism in particular, in complete opposition to the Christian faith. This analysis is a central element in the lecture by Manuel Pizarro, one of the leaders of the movement, for whom the Darwinian emphasis on extinction as the inevitable complement to evolution by natural selection strikes at the heart of the Church's social and political agency, as expressed through its charitable organizations. A society devoid of altruism is one divorced from the Catholic ideal of care for the meek. According to Manuel Pizarro, Darwin's work could be read as a call to abandon the weak and downtrodden.

> Bear in mind that there live among us many creatures that resemble man only in the shape of their bodies, whom nature has condemned, by virtue of the original inferiority of their lineage, to lose the battles of the struggle for existence, and thus to irremediable disappearance. If you seek to help these creatures, contrary to the laws of nature, know that you are bringing about disastrous decadence of your own people. By preserving pointless existence, you bequeath to your descendents the inevitability of a true barbarian invasion, its brutal hordes bred, with your assistance, on the very soil of your country.
> Such is the charity of the modern-day philanthropist!
> Such is the morality to which our century pretends!
> Such is the *good news* that liberalism proclaims to the nations of the future!
> (Asociacion Católica de Buenos Aires 1885, 47; emphasis in original)

Pizarro attributes the first paragraph of this text to Darwin's *Descent of Man*, in which we have found no obvious source for it. His lecture also cites Herbert Spencer. It appears likely that Pizarro has combined and glossed aspects of the work of both thinkers in the service of his primary objective, to remind his listeners of the importance of altruism and the love of one's neighbor. Liberalism, by contrast, is a "strange philosophy" whose twisted parody of Christianity proclaims liberty, equality, and fraternity in politics, while declaring, with Darwin, the "rule of force" in morality and "inscribing in our hearts the cold, sad, and cruel maxim, *fatalism and egoism!*" (48; emphasis in original).

It is important to note once again that, despite his rhetorical excesses, neither Pizarro nor any of the other speakers at this Catholic Assembly is really railing against science per se, or even against evolution as a scientific hypothesis. They are chiefly concerned that, in the course of the transformation of the moral principles on which the social power of the Church rested, the Church would come to be seen as an obstacle to progress, rather than its ally. Pizarro and other assembly participants champion both science and progress—as tempered by a Christian perspective. According to this view, the gospel itself proclaims "the evolution of souls toward saintliness, justice, and liberty."

> This is the true *struggle for existence*; only the existence in question is *eternal life* and the struggle one of *charity*, *love*, and *fraternity*, in which the *superior classes* neither sacrifice the *inferior* nor allow them to perish, instead caring for them and helping them until, invigorated and strengthened by time, they attain *Life*.
>
> This is the law of evolution by *mystical selection* [*selecciones místicas*], realized in time. (54; emphasis in original)

In the true struggle for existence, the gospel of Jesus Christ serves as a guide. It may be understood by strict analogy to Darwinian evolution, with roles for struggle and competition and, by divine grace, adaptation. Properly catholicized, evolutionary discourse need not be rejected.

Argentina's leading Catholic intellectual, José Manuel de Estrada, gave the assembly's closing address, which we discussed in the previous chapter. In his view, "man . . . either ascends toward angelic beatitude by restraining his passions and gazing toward the heavens, or by denying the heavens and releasing his passions, he descends into bestiality. Human genealogy leads to the latter path, according to the crazy naturalist philosophy" (476). The "crazy naturalist philosophy" in question was Darwinism. In consequence of its widespread adoption, "the silence of servitude and abjection [prevails], provincial governments are oppressed or oppressive, and everywhere we find the struggle for life, a conflagration of appetite and concupiscence!" (477). In the absence of any hope of transcendence of the sort traditionally provided by religion, the negativity of evolutionary thought could only degrade the people.

Whether their fears were justified or not, the Catholic opposition's perception of the extent to which the evolutionary worldview had penetrated every aspect of society was quite accurate. At the end of the century, Carlos Octavio Bunge published a book of readings intended for use in fifth- and sixth-grade classrooms. In its twentieth edition, published in 1910, we find a poem on the history and traditions of the Argentine people entitled "Ofrenda a la Patria" (Offering to the Fatherland). Nine- and ten-year-old children were expected to recite a pledge whose terms flowed directly from Darwinian logic.

> 1. By my God and my blood
> I offer you my life;
> What I am and what I have
> I owe to you, my Fatherland.
>
> 2. All I sing and all I dream—
> The brimming chalice of my life
> I pledge to you, my Fatherland
> Before your heroes' altar.
>
> 3. I'm not frightened of the battles
> In the struggle for existence,
> For I know that you, my Fatherland
> Will always be victorious.
>
> 4. And if, in this great struggle
> I should draw my final breath,
> With my last gasp I will shout,
> "Live and triumph, Fatherland!"
>
> 5. What I am and what I have,
> I owe to you, my Fatherland:
> I have offered you my life,
> May you use it in your service. (C. Bunge 1910, 1)

Clearly expressed in the fourth stanza of this poem is the notion that the struggle for existence is not a matter of mere individual posterity. The culture of extinction demands of individual citizens the willingness to sacrifice the hope of such posterity in order to ensure the continuity of the nation. Those destined to biological extinction may console themselves with the hope that the nation, at least, will live on. Nor is this a marginal text:

Bunge's book had been officially adopted for use in the primary school curriculum. So widely was it used, its reputation eventually reached as far as the United States, where excerpts from Bunge's work were published in an anthology of Argentine sources edited by Garibaldi and Cincinato Laguardia. They call Bunge "a very modern Argentine writer" and cite, among other works, *Nuestra Patria*, "an anthology for use in the Argentine schools, containing, besides extracts from other Argentine authors, many episodes written by Bunge himself" (Laguardia and Laguardia 1919, lii). Nor was the introduction of evolutionary principles confined to the officialist rhetoric of primary education. A 1907 report on the state of agricultural education in the province of Buenos Aires notes that instruction in "reading, calligraphy, revision, arithmetic and geometry also has a role to play in agricultural studies, as an auxiliary to the acquisition of skills necessary for superior performance in the struggle for life" (Godoy 1907, 54).

Spencer, Darwin, and Haeckel formed the basis of many university curricula, especially in the natural and social sciences. José Ingenieros, of whom we will have a great deal more to say later, integrated the main elements of Darwinism in his broadly socialist approach to the psychology curriculum. The theory of evolution by natural selection proved extraordinarily flexible in its capacity to contextualize a wide range of political ideologies. Though Ingenieros emphasized the fraternity of human beings and the universality of their rights, he also insisted upon extinction as a necessary condition of human evolutionary progress. In 1905, while en route to Europe, he had the opportunity to observe the inhabitants of the Cape Verde Islands, whom he immediately associated with the population on whom the Spaniards had drawn for their New World slaves. Classic pre-Darwinian Argentine narratives had treated them as exemplary victims of Spanish oppression, as Ingenieros recalls; "On reading Mitre or López, for example—and here I confine myself to the most important of many such examples—on the importation of African slaves to the old Spanish colonies, we find ourselves imagining them as victims of White iniquity, and sympathizing with their pain" (Ingenieros 1908c, 269). The modern reader is led "to think of these African slaves as not unlike today's Negroes . . . in their colorful costumes." But such sympathy is "a grievous error, one that falsifies any interpretation of the historical role played by the Negro race in the formation of the American character and the American people." Before their improvement by their Argentine sojourn, the original slaves must have been "much more like those who now people San Vicente: the most loathsome residue of the human species." Slavery is justifiable on purely beneficent grounds: "Our civil law displays similar generosity in sanctioning the tutelage of the feeble-minded, the asylum of the deranged, and the protection of animals. Slavery is nothing but the political and legal sanction of a purely biological reality" (269).

While such racism might appear inconsistent with the socialist ideology of fraternity Ingenieros so vigorously defended, sensible political ideals must, he argues, be accommodated to biological reality. "Human solidarity" in ignorance of that reality can only be "a lyrical, irrational conceit. The 'rights of man' apply legitimately only to those who have reached the same stage of biological evolution." And not all humans are equally evolved. "Men of the white races, even those belonging to more inferior ethnic groups, are separated by a wide gulf from these other beings, who bear far more resemblance to the anthropoid apes than to civilized Whites." There can be no place for these inferiors in civilized life, except, perhaps, as museum pieces, their crania displayed side-by-side with mounted fossils of extinct animals, for "in anthropological terms, they are nearly simian—even more so than the plates in anthropology texts, or the collections of skulls in museums, would lead one to believe" (272).

Short of his actively militating for the return of slavery, it is difficult to imagine Ingenieros taking a position more at odds with his generally egalitarian views. "Men of the colored races," he baldly asserts, "ought not to be our equals, in politics or in law, for they are unsuited to the exercise of civil duties, and thus cannot be considered 'persons' in the legal sense of the word." To be sure, "each rule has a thousand exceptions, and with this claim as with other assertions about social phenomena, it would be a mistake to treat a relative truth as an absolute." It seems not to have occurred to Ingenieros during his visit to Cape Verde that the ground for the distinction between such exceptions and the general rule might turn, not on biology, but on access to education or other scarce cultural resources. The majority of his fellow Italian immigrants were doubtless as illiterate and uncultured as the Cape Verdeans—but only the latter were marked for extinction. "Men such as this cannot survive in the struggle for life. Natural selection, as inevitable over the long term for man as it is for other animal species, eliminates them on every encounter with the white races" (272).

Far from reflecting on the social and economic degradation of the Africans he had encountered, Ingenieros is convinced he has observed their race at its very best, "for centuries of contact with Whites have allowed only the most elite specimens to persist. Similarly, the Negroes still found in the Americas represent the very flower of those introduced by the Spaniards, adapted as they are to the conditions of life in our Europeanized environment." Viewed in the cold light of scientific objectivity, any effort to come to their defense must be seen as both antiscientific and potentially harmful to the national interest.

> To campaign in support of the inferior races is to be antiscientific. The most we might do is to protect them in such a way as to allow for their gentle extinction, facilitating the provisional adaptation of those few who are capable of it. These discarded bits of human flesh are to be regarded with pity. They should be treated well, at least as well as the centenarian tortoises of the London Zoo, or the trained ostriches of Antwerp. We would not vote with the draconian jury of the poetically named Mississippi town of Magnolia, who recently sentenced Theresa Perkins, a white woman, to ten years of forced labor for the crime of having married a Negro. But it would be absurd to advocate the indefinite preservation of this race, as it would be to favor the crossing of Negroes with Whites. (Ingenieros 1908c, 274)

Ingenieros concludes his discussion by drawing connections between his quasi-Darwinian understanding of racial inferiority and familiar socialist discourse. Because the "constitution of a society inevitably reflects its economic roots," and the "material conditions of a people correspond to the particular aptitudes it brings to bear in the struggle for life, and in adapting to its environment," it follows that the people "of the Cape Verde Archipelago, and others like it, are doomed to misery" (274).

Far from being unique, the merciless tone taken by Ingenieros is actually representative of an Argentine intellectual community that had embraced, as the Catholic opposition charged, a rather extreme culture of death—death without hope of resurrection. As evidence of the extent of the cultural preoccupation with extinction, the case of the treatment of a young Indian woman named Damiana merits close inspection. In 1898 F. Lahille, director of the Department of Zoology at the Museo de la Plata, published a study of the Guayaquí Indians of the upper Paraná, in the northeast of the country. Lahille describes a confrontation between a group of Guayaquí and a group of ranchers, in which all of the former were killed, save for "little Damiana. She was a year old, more or less, when she was taken in by those who had murdered her parents" (Lahille 1898, 454). Referring back to an original account of her capture, Lahille draws inferences concerning the Guayaquí language, noting that the infant Damiana had apparently repeated the words for "mother," "suck," and "want," indicating that she "was asking to be nursed by her mother" (456).[4] We are told little more, though it is clear that Damiana remains an object of considerable anthropological interest.

In 1908 Robert Lehmann-Nitsche, a German anthropologist active in Argentina and president of the 1910 Congress of Americanists, takes up the story of Damiana again, beginning with her early history as the sole survivor of an attack by settlers of the Sandoa Ranch, near Encarnación, on a group of Indians. The little girl, "called Damiana, after the saints-day of the massacre" (Lehmann-Nitsche 1908, 91), was taken in by settlers apparently acquainted with the explorer de la Hitte, whose description of the massacre is cited. Another well-known anthropologist, Herman ten Kate, documented this rare specimen of the reclusive Guayaquí people, taking photographs and performing a standard anthropometric battery. In 1898 Damiana was removed to San Vicente, in the province of Buenos Aires, where she lived with the mother of Alejandro Korn, a prominent physician and director of the Melchor Romero hospital.

In San Vicente Damiana was installed as a maidservant and "developed normally." But with the onset of puberty, her "sexual libido became manifest in such an alarming manner that no amount of education and chastisement on the part of her host family proved effective." She considered sexual acts to be "the most natural thing in the world, and set out to satisfy her desires with the instinctive spontaneity of an innocent being." She was driven out of the San Vicente house in 1907, and Dr. Korn employed her as a hospital orderly at Melchor Romero, in the province of Buenos Aires, while awaiting the opportunity to enroll her in a corrective institution. It was there that Korn gave Lehmann-Nitsche permission to complete a new set of photographs and measurements. He was able to complete this task, but "two-and-a-half months later the unfortunate Indian died of raging consumption" (93), suffering the fate seemingly foreseen for her, if delayed, from the moment of her capture.

We have reproduced Lehmann-Nitsche's photograph, in which Damiana appears nude, in a pose typical for anthropological photographs of aboriginal women—despite her guardians' concern regarding her apparent shamelessness. Lehmann-Nitsche takes special note of her facial expression. "I treated her twice, and both times found her reserved, shy, and mistrustful. This is evident from the peculiar expression of her gaze (see photograph). I was roundly startled to hear her speak German, a language

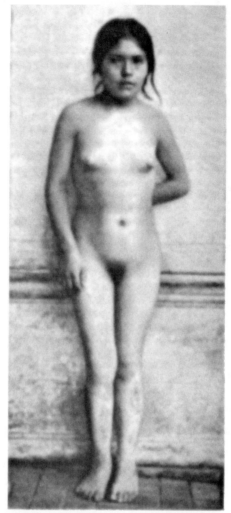

India Guayaqui de 14 á 15 años, según fotografía del Dr. Lehmann-Nitsche

FIGURE 3. From Lehmann-Nitsche 1908, 93.

which she had learned in San Vicente and spoke quite well, though with a degree of accent and slang useage derived from Spanish, which was quite naturally her primary language" (93).

Though noting his own surprise, Lehmann-Nitsche draws no inferences from Damiana's remarkable proficiency in German. But he has a great deal to say about Damiana's anatomy, measurements of which he compares in detail with data from German girls of the same age. He is also perfectly prepared to draw conclusions about his subject's various capacities on the basis of these measurements, noting, for example, that "the little Indian's frontal region, the seat of intelligence, was very well developed" (98). No amount of intelligence could save her, however, from remaining essentially a specimen, a status to which the members of near-extinct groups were invariably consigned. "The little Indian's head, with her brain, was sent to Professor Juan [Hans] Virchow of Berlin, for the study of her facial musculature, brain, etc. The cranium was opened in my absence, and the incision placed too low. The preparation of the orbital musculature Professor Virchow had hoped for will thus prove impossible, but the brain has been most admirably preserved. The head was presented to the Berlin Anthropological Society ... and we may soon expect further publications" (98). At the beginning of his article, Lehmann-Nitsche refers to Damiana by name. His is not, after all, the first publication on this particular individual, and his use of the subject's name is doubtless intended to help his readers bring these prior studies to mind. But having recalled the particulars of her history, her name is no longer necessary; she becomes simply a member of a soon-to-be-extinct type, a "little Indian" (*indiecita*). Following the arrival in Berlin of the formalin-filled jar containing the preserved sections of her head, Hans Virchow (son of the celebrated Rudolf Virchow) published a brief notice concerning this rarity in the *Zeitschrift für Ethnologie* (Virchow 1908, 117–20). There is, of course, no mention of the specimen's name. Considerable attention is given to the unfortunate placement of the facial incision, it having destroyed the musculature in which, Virchow believes, diagnostic racial features are most plainly manifest. While acknowledging the value of Lehmann-Nitsche's specimen ("to present the head of a Guayaki Indian in Europe is a very special opportunity" [Virchow 1908, 117]), Virchow in fact devotes nearly the entirety of his article to lamenting its inadequate preparation. Where Lehmann-Nitsche notes her mistrustful expression, Virchow can only complain that tight packing has squashed her nose.

To a Hans Virchow, in distant Europe, clinical detachment toward his specimen may have come naturally. The head reached Berlin after traveling a long way in space and, one senses, a long way in time as well. Damiana's head was prized as an exemplar of true "stone-age man," as the Guayaquí had come to be known in the literature of the time, living fossils, in effect (see, e.g., Steinen 1892, 248–49; Ehrenreich 1898, 73–78; Vogt 1903).[5] But to those who lived among them, the fact that primitive humans who were

Fig. 1.

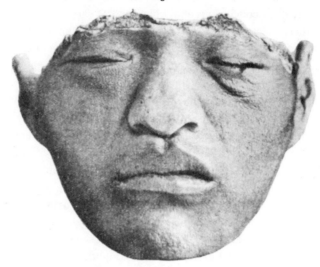

Kopf eines Guajaki-Mädchens (von vorn).

FIGURE 4A. From Virchow 1908, 118.

Fig. 2.

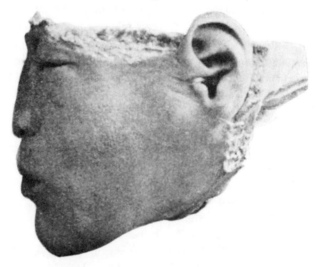

Kopf eines Guajaki-Mädchens (von der Seite).

FIGURE 4B. From Virchow 1908, 119.

supposed to be extinct remained very much alive made this degree of detachment impossible. In consequence, in Argentine discourse, "extinct," when applied to the groups considered in this chapter, frequently became less of a descriptive term, than a normative one.

In 1903 Ernesto Quesada (1903) comments on the present status of evolutionary thought in an essay subtitled "The Struggle for Life and Rest" (La lucha por la vida y el descanso),[6] an extended reflection on a novel published that same year, *Reposo* (Repose), by Spanish historian Rafael Altamira. Quesada argues that the duality of evolution and extinction has serious and unacceptable consequences in societies perceived as governed less by the promise of the former than by the threat of the latter, with its logic of exclusion and discontinuity. Altamira's novel is "one of those works . . . that exposes a human problem, probes it in all directions, cuts mercilessly into living flesh and, sustained by ironclad logic, unmoved by the illusions it shatters and the incurable wounds it opens, imposes a cruel, cold, and indifferent philosophy as the inevitable solution" (E. Quesada 1903, 262). Yet in its very pitilessness, Quesada asserts, the novel reveals the intrinsic shortcomings of its own Darwinian discourse, in which the laws of nature entail that "only the strong, the vigorous, the resolute triumph, while the weak and the less determined succumb. . . . Little by little, thanks to the natural elimination that indubitably occurs over the course of this unending clash, progress is accentuated, because only the most dedicated or the most audacious prevail, with all those falling by the wayside who, whatever their personal qualities or intellectual training, never learned to take up arms and fight, or lacked the will to persevere" (266). In the midst of this struggle for life, "humanity appears to be compelled toward the dangers of a colossal dance of the dead, its dancers all feverish, punching one another right and left, almost without noticing whom they hit, all blindly pushing and shoving to reach their goal. . . . Life is struggle, life is war" (267). But to become cognizant of this condition is to realize the "apparent futility of strictly doing one's job . . . [given the] holocaust in progress" (260). The greatest problems faced by modern humankind are the challenge of adapting to the complexities of modern life and the prospect of falling into degeneracy and failure. Educated persons aware of the "dance of the dead" in which they are engaged find themselves "breathless"; indeed, Quesada reports reading Altamira's novel on an Easter holiday spent in the country, to which he has fled to escape "the terrible specter of neurasthenia" (257). This specter haunted many, perhaps most of the Argentine intellectuals of the late nineteenth century, who were constantly second-guessing themselves, ever alert to the signs of degeneracy, prelude to extinction. Quesada's obsession was shared by Ingenieros; Ameghino suffered a nervous breakdown; and Bunge was forever on the verge of one.

When neurasthenia was understood as an urban symptom of the failure to adapt to the modern conditions of the struggle for existence, the desire to avoid it quite naturally led some, including Quesada, to search for flaws in the Darwinian logic in terms of which this condition was defined. And so, Quesada charges, "in invoking the famous Darwinian law," Altamira overlooks the fact that the "noisy laws of competition and natural selection are not so absolute as they appear at first glance. . . . Study history, analyze life, and it becomes plain that, fortunately, the strongest do not always triumph. To the contrary, it is easy to show that when, in a given isolated instance, they do prevail over the weak, the clear, reciprocal response is for the defeated to bond together in turn, to challenge and undo their conquest and, in short order, to defeat and annihilate the conqueror. Sheep may also achieve victory, even against millions of wolves" (268). Having challenged the received notion of the struggle for life, Quesada finds his vision of nature transformed. Evoking pre-Darwinian images, he extols nature's "commanding beauty and Olympic serenity. The sky, the sun, the flowers: all conspire to live, and live placidly" (276). If contemporary Darwinian materialism cannot capture this harmony, it is in need of correction.

Quesada attempts to find an equilibrium between the theories he had long defended and some of their less palatable social consequences. Among the young writers of this period, we find a reaction against the negativity embodied in the emphasis on extinction as the necessary by-product of evolutionary law. The constant struggle of modern urban life must always be balanced by the periodic return to nature—pastoral nature, not "nature red in tooth and claw." Quesada's nature is no longer an implacable judge, selecting some while discarding others. Instead, he extols the "beneficent interval, powerful restorer, most noble therapy of nature! When the aggravation of struggle has eased, and the soul is made serene, man sees his own defects, or the injustices he may unwittingly have committed, with much more clarity. He nobly concedes his errors with regard to doctrines, people, or things, and when he returns to the fray, wholly restored, he does so with more equanimity of spirit, more respect for others, and greater inclination to generosity, indulgence, and the forgiveness of past slights" (319). Quesada's text clearly displays the misgivings experienced by evolutionary thinkers at the beginning of the twentieth century, as they came to grips with the ways in which their own peripheral status undercut their ability to embrace Darwinian triumphalism.

Natural selection has a different meaning to those comfortable in the knowledge of being selected than for those living in fear of degeneracy and extinction. The latter thus sought to correct Darwin in ways that dovetailed with various aspects of pre-Darwinian thought. To hold fast to the hope of future continuity in the face of the Darwinian threat of discontinuity required a reexamination of the continuity of the colonial past with the Republican present, though Rosas's liberal opponents had consciously erased this past. Restoring broken lineages demanded the reconstruction of the national pedigree, to be accomplished only by the rediscovery of colonial history. Quesada thus calls for the reconstitution, "piece by piece, of three long centuries of memorable past events, three centuries crying, impatiently, for some audacious explorer to take their virginity" (327). It is no coincidence that Quesada chooses his discussion of a book by Altamira in which to make his plea, for Altamira was a leader of the Spanish intellectuals calling for the resumption of cultural exchange with the former colonies, with a view toward rediscovering the intellectual common ground of the Spanish-speaking peoples.

In the same year in which Quesada was voicing his misgivings regarding the logic of natural selection and its received interpretation, Ricardo Rojas published a collection of poems entitled *La victoria del hombre* (The Victory of Man). Within a few years, this writer would be acknowledged as a leading nationalist thinker, spearheading a movement that called for a cultural revolution to overthrow the materialist doctrines responsible for the anxieties discussed in Quesada's article. From the following three stanzas of "La tierra prometida" (The Promised Land) we may gain a clear sense of the ideas that had become important to many young intellectuals of the day.

The Precursors
Behold the shadow of these souls—
Around them, all turns into glory
Their feet tread the same mud as ours,
But their visages are bathed by stars:

Darwin's gaze is fixed upon the past,
Absorbed by life in process,

> When suddenly a cell surprises
> With the mysterious ascent of progress
>
> Renán, apostle of the new life
> Leaves convent crypts behind, and soars—
> His mind, become a daring eagle,
> Describes the curve of his immortal flight
>
> Marx forges, in the crucible of his genius
> The yearning for an egalitarian world
> On whose titanic stage he shatters
> The brazen chains of wages. (Rojas 1903, 39)

In the following stanzas, we find also mention of such figures as Dostoevsky, Ibsen, and Wagner—but the place Rojas accords to Darwin is significant. His is one of the glorious souls, but his "gaze is fixed upon the past." Paradoxically, the new science was not moving the country forward, as it had been in the mid-nineteenth century, but forcing intellectuals to reconnect with what had been. The return to genealogical thinking is one of the most relevant consequences of Darwinism in Argentine culture.

Florentino Ameghino and Francisco Moreno: Documenting Extinction

One of the best metaphors for the uncertainty felt over the necessity of personal disappearance in service to the future can be found in a work written by the man who best represented scientific evolutionary theory in Argentina, Florentino Ameghino (1854–1911). Unlike his close contemporary Francisco P. Moreno (1852–1919), Ameghino did not belong to a distinguished creole lineage and thus lacked the former's entitlement to the status of quasi-official patriotic naturalist, charged with classifying and organizing the nation's past. In a world in which the naturalists were almost all part of an elite that controlled politics and culture, Ameghino was largely alienated from his colleagues, with whom he would have difficult relationships all his life.

By all accounts his youth was nothing short of prodigious (see, e.g., Cabrera 1944, 12ff). Shortly after completing his studies at the Luján municipal primary school at 13, he was appointed assistant teacher. A year later he attended Normal School (Escuela Normal de Preceptores) in Buenos Aires, racing through the program and gaining a position as assistant principal for primary education in Mercedes, becoming principal at the age of 16. By 1871 he had read Charles Lyell and Hermann Burmeister, and probably Charles Darwin as well. Ameghino's interest in paleontology and paleoanthropology dates back to his childhood in fossiliferous Luján and abided throughout his studies and early career in education. By 1875 he had completed the manuscript of *La antigüedad del hombre en el Plata* (On the Antiquity of Man in the Plata Region) (published in 1880) and begun his *Diario de un naturalista* (A Naturalist's Journal), inspired by Darwin. That same year he also penned his "Nota sobre algunos fósiles nuevos de la formación pampeana" (Notes on Some New Fossils of the Pampan Formation). Notoriously, in this paper as in *La antigüedad del hombre en el Plata*, he argued for the Pampas as the site of earliest human evolution.

In 1889 Ameghino delivered a lecture on his *Filogenia* (Phylogeny), published in 1883. *Filogenia* had presented a mathematical system for deducing phylogenetic relationships on the basis of comparative quantitative morphology. The 1889 lecture "Vision y realidad" (Vision and Reality) is advertised as an "allegory" for this phylogenetic method. As one might expect, the technical subtleties of Ameghino's theory are absent. Of particular interest is Ameghino's account of how this theory came to be. His years of painstaking research had led, not to a measured, rational conclusion, but to a sudden vision in which "all the knowledge I had taken so long to acquire merged in a whirlwind of memory" (1889, 172). Before proceeding to describe the vision in detail, he offers the following summary.

> The entire natural system, of which I had caught distant, misty glimpses so many times before, now appeared before my eyes in all its immense, humbling, sublime glory. A thousand hands would not have sufficed to take down on paper the innumerable, fleeting images that passed through my head in that brief instant, with unfathomable, tumultuous speed, like a vertiginous avalanche or the gyrations of some magic kaleidoscope. My constitution could not withstand so fierce an onslaught of thought. My strength failed, and I broke into a cold sweat, followed by general lassitude and a sort of vertigo in which, by involuntary reflex, my brain compiled a fantastic synthesis of that wave of ideas, fierce as the ocean surf, a synthesis that impressed itself indelibly on my memory, and with such intensity that its mere evocation is enough to make me swoon. Oh! I can still see that ferocious, colossal scythe at work! (173–74)

Retracing his steps, Ameghino next describes having become "an inhabitant of interplanetary space," from which vantage he could see the newly formed earth "shrouded in an atmosphere charged with carbonic acid and aqueous vapors" (174). On the "lowest, most tranquil surface" of the young orb, there appears "something that was neither earth nor water, neither solid nor liquid: a thick, lumpy, slimy, semi-liquid substance, capable of rapid, protracted movement toward a common center, which for its part slowly took shape, finally forming an amorphous, but stable agglomeration."

> This colloidal mass of animated matter, in continual motion like the waves of a furious ocean buffeted by storm, obeyed a powerful centripetal force, drawing it in toward the center, which rose, bit by bit, to form a column of extraordinary girth and height, splitting at its top into several branches. These continued to rise, gradually diverging. Suddenly, the column shuddered at its base, collapsing, melting into a vast, gelatinous sheet, in which the higher branches were left planted, separated from one another but intact and alive. These continued their independent growth, themselves becoming columns, or gigantic trunks, and in turn giving rise to a considerable number of branches bearing secondary twigs. (174–75)

Finally there appeared in the distance "a sort of crescent: it was a giant scythe, held by an invisible, but powerful hand. It advanced, resolutely, and with a formidable stroke severed the tips of the trunks and branches, which fell to the ground." The scythe proceeded to harrow the trunks, "until they began progressively to fall, in sections, the decomposing remnants of which accumulated over the centuries in layers of dust, one on top of the other, sustaining and nourishing the separate crowns" (177).

> These continued to grow and reproduce until they had filled all available space with an immense, dense forest. And now there was too little food, light, heat, moisture, air, in short, too little space to support so much life, and the great struggle for existence began. While some attained colossal proportions, others became rachitic. . . . Regardless, as seen from above, the forest as a whole never stopped growing, becoming ever stronger, more stately, more robust and vigorous, more splendid and beautiful, ever sprouting new branches, each with its innumerable branchlets, twigs, buds, and leaves . . . while the terrible, ferocious scythe hewed at the very bases of the trunks, severing them for all eternity from the bonds which link us with our ancestors. (177–78)

Growing frightened of the scythe, the interplanetary voyager leaves for an eon, returning to find that "the atmosphere, now relatively calm, had become clear and transparent, while the Earth's crust, apparently at rest, had taken on definite contours. In grand temples consecrated to the study of nature, successive generations devoted themselves to the task of cataloguing all life forms both living and extinct, describing them in minute detail" (178). These researchers "had probed every character, even the most insignificant of them, without extracting the magic word with which they might lift the impenetrable veil that hid from them the grandiose structure of the chain of being" (179). The naturalists toiled for centuries, until the Ameghinos among them, "building on the vast collection of observations their predecessors had amassed, concerned themselves with reducing the distinctive characters of life forms to fixed formulas, assigning a numerical value to every anatomical factor, comparing these numbers with each other by means of a succession of subtractions and additions, and finally placing them in natural sequences whose extended branches diverged in the future, converging in the past" (179). Returning to consciousness, Ameghino notes that his fantastic vision had shown him the realization of his great scientific goal, to re-create "the lines of descent between life forms" (182).

When European followers of Darwin applied the theory of evolution by natural selection to the social realm, they often found themselves celebrating the triumph of their own favored race. In Argentina, as we have repeatedly shown, matters were different. Evolution meant death. Ameghino's vivid image of an invisible hand wielding a scythe recalls Adam Smith's metaphor for the market. In Ameghino's account, the naturalist neither guides nor preserves the stronger branches, but must instead be content to reconstruct the lineage, proving that the parts no longer there, severed by the ferocious scythe of death, belonged to the same trunk. This vision is a startling representation of the hopes and fears of Argentina's intellectual and scientific class, recalling as it does the difficulties inherent in the task of building a civilized identity in America. Like Quesada (see above) and Moreno (see below), Ameghino sees his own role in these efforts as bent toward the reconstruction of lineages. Francisco Sicardi's *Libro extraño* (Strange Book), published between 1894 and 1902, provides a good literary counterpart to Ameghino's understanding of his scientific mission. Sicardi, a medical doctor, adds a Nietzschean tone, typical of the time, to his interpretation of the culture of death. "Because facts need sites, dates, and creatures, I write these chapters of the book, this book that carries within itself the seed of death, because in art, only the superhuman has lasting life, reflections of the truth for all time and in every place.... It doesn't matter what happens next; let us write. I know that the grave awaits with its marble lid lifted and ready always to fall" (Sicardi 1910, 4).

Francisco Moreno, founder of La Plata's museum, adds another interesting example that illuminates this interest in dead bodies. He had a lifelong fascination with Indian remains. By the age of 21, when he undertook an early expedition to the south of the province of Buenos Aires, he had already begun to collect the crania and skeletons with which, as a leading anthropologist, public figure, and museum director, he would later seek to document the extinct races of the Republic: "I reaped an abundant harvest of crania and skeletons in the cemeteries of the indigenous peoples who once lived in the immediate vicinity of Azul and Olavarría.... That endless plain, with its great mysteries, impressed me so greatly, and attracted me so powerfully, that on my return to Azul I resolved to penetrate into the soil, to discover what it held" (Moreno 1893, 66–67; qtd. in Stagnaro 1993, 59). Moreno's obsession with profaning Indian cemeteries features prominently in the letters he wrote to his father during his Patagonian expedition of 1875. The letter of April 5 carefully describes the crania he has collected so far and those he still hopes to obtain. Significantly, throughout these descriptions, Moreno draws no distinction between his deceased donors and his living prospective donors. As for the Indian populations from which they were drawn, he has no difficulty speaking of them as already extinct; as he understands it, his own task, that of preserving their remains so as to document the progress of national evolution, is thus essentially the same as that of the paleontologist.

> Although I doubt I can obtain the number of skulls I had wished for, I am certain that by tomorrow I will have 70. Today I shipped you a crate containing 17 of them, which I urge you to send for as soon as possible, as the courier remains unaware of the sort of merchandise he is carrying. Any other time I might have met my goal, but now, with Indians in such disarray, it is impossible.
>
> I expect it will not be too long before I acquire the bones of the entire Catriel family. I already have the skull of the illustrious Cipriano, and the whole skeleton of his wife Margarita. Now it appears that his younger brother Marcelino, the leader of the present uprising, hasn't long to live, either. Yesterday in the Nievas arroyo he surrendered to [Marcos] Levalle's Remingtons, and his dear brother Juan José, who turned over Cipriano, has promised to do the same with Marcelino. I have sat at a meeting with Juan José, who strikes me as a most vulgar and sly Indian.

I read with pleasure the article by Zeballos on the murder of Cipriano Catriel, and have had occasion to consult here with people who knew him, all of whom agree with what Estanislao [Zeballos] has to say on the subject. The head (Catriel's, that is) remains here with me. I examined it a while ago, but even after a bit of cleaning, it continues to stink. It will travel with me to the Tandil, as I am unwilling to separate myself from this jewel, for which I am greatly envied. (Moreno and Moreno 1942, n.p.)

The urgency with which Moreno conducted his collection activities was born of his strong sense that the rapid disappearance of Indian tribes had gone largely undocumented. This disappearance, coupled with the rise of the modern nation, demanded the creation of museums in which the process by which the contemporary Argentine population had evolved would be portrayed didactically. Evolution could only be established and understood by reference to extinction.

As we have discussed, in the performance of his chosen task, Moreno always saw himself as bound, to use Sarmiento's phrase, to "follow in Darwin's footsteps, continuing where he left off, or taking instruction from his insights on later discoveries" (Sarmiento 1900c, 177). In his 1879 account of his 1876–1877 Patagonian expedition, he follows the trajectory described in Darwin's *Journal of Researches* with almost religious zeal, here and there providing further detail on the English naturalist's observations, or drawing attention to something he had missed. At one point, rather than provide his own description, he simply cites the *Journal*, for "Darwin describes [these matters] so precisely that I prefer to reproduce the paragraph he devotes to them in his journal, rather than providing one in my own arid style" (Moreno 1879, 140). Reporting on a layered sedimentary formation that Darwin had missed, Moreno is quick to note "what immense interest it would have held for the illustrious naturalist!"

> The history of past generations lies buried in the sandy bowels of this grey cliff-face. Within the successive layers we find preserved the remains of the beings nature deposited in this bed, some of them whole, others in tiny fragments, testifying to the other, later organisms generated in the indefatigable exercise of her powers, on the genealogy of those that preceded them upon the stage of life, preparing the way for their own entrance. These animals, whose remains roiled the oceans and rivers before being deposited beneath the surface of Patagonia, display the great richness and variety of the beings that once paraded their curious forms across the tertiary landscapes. (303)

In this same book we also find mention of Moreno's quest for indigenous crania, though the matter is treated with somewhat more discretion than in his private letters. Chapter 8 relates the story of one San Slick, "a good Tehuelche" he had met on a previous expedition. Reuniting with his friend, Moreno is surprised to find him reluctant to submit to anthropometric measurements, though he consents to be photographed. On a later encounter during the same expedition, Moreno discovers that, "although we were still friends, he would not allow me to get close to him." A year later, San refuses an invitation to accompany him, worried that Moreno "wanted his head." But such suspicion and caution notwithstanding, his fate is inescapable.

> This was his destiny [to bequeath his head]. Days after my departure, he set off for Chubut, where he was murdered by two other Indians in a night of debauchery. On my return, I learned of his ignominious end, enquired into the location of his grave and, on a moonlit night, exhumed his cadaver, whose skeleton is now preserved in the Buenos Aires Anthropological Museum. This sacrilege was committed in the service of the osteological study of the Tehuelches.
>
> I did the same with the skeletons of chief *Sapo* and his wife, who had died years earlier while their clan was encamped there. Both had been buried in a Christian cemetery, though the indigenous practice of placing the cadavers in sitting position had been observed. (93)

The fatalism with which Moreno describes San Slick's inevitable destiny gives us some sense of how he viewed his friend. He made no distinctions among the Indians still walking, those in fresh graves, and those in ancient cemeteries. All were victims of the same process of extinction, to which witness must be borne. The crania of Indians were national treasures, for their disappearance revealed the forces of evolution in action. Even Christian cemeteries were no refuge. Though such places constituted one of the new marks of civilization upon the landscape, the preservation of genealogical evidence was more important than any sacrilege implied in their violation.

Moreno's view of the indigenous population, its destiny, and the duty to preserve evidence of its vanishing race, would become canonical. So would his view of Patagonia. His account of his expeditions would become the most important description of the region, next to Darwin's, and would be consulted by naturalists the world over. Even before the publication of his journal, his expeditions attracted widespread attention, as witnessed by the publication of his "Description des cimentères et paraderos préhistoriques de Patagonie" (Description of the Prehistoric Cemeteries and Paraderos of Patagonia) in the *Revue d'Anthropologie* (discussed in our introduction) and an 1877 notice in the *Revue de Deux Mondes*. An article entitled "Les dernières explorations dans La Pampa et La Patagonie" (Recent Explorations in the Pampa and Patagonia) reviews the most important publications on the subject, including works by Darwin, Burmeister, and Moreno, whose excavations of Indian cemeteries are discussed in great detail (E. Daireaux 1877).

As discussed above (in the introduction and chap. 1), Moreno's ultimate goal was to establish a national museum devoted to the evolutionary history of Argentina. Even his early, private efforts attracted considerable attention. Following a visit in 1887, American naturalist Henry A. Ward described the Museo de la Plata—chartered by the national government in 1882, anchored by the donation of Moreno's vast private collection, and directed by Moreno himself—as "one of the ten finest museums in the world" (H. Ward 1890–1891).Ward's article is full of admiration for Argentina and in particular for the city of La Plata, new capital of the province of Buenos Aires, built in its entirety following the federalization of the city of Buenos Aires as the capital of the nation in 1880. The museum itself, cornerstone of the new city, is a "splendid" example of "beautiful Greek architecture," festooned with life-size busts of such savants as Aristotle, Lucretius, Linnaeus, Lamarck, Cuvier, Humboldt, Darwin, Owen, and Broca. Its central rotunda features frescos "representing scenes from what is now the Argentine Republic, in past geological epochs. These paintings realistically depict primitive man, the prehistoric Argentine, in the company of animals of the quaternary period. Daring though it might seem, this composition is really nothing but an exposition of the facts acquired by science" (147–48). Ward also discerns, correctly, that the museum is intended to serve not only science, but the nation, by paying homage to genealogies lost in the course of its formation. This is a museum of extinction, depicting past sacrifices in the struggle for life as the necessary antecedents of the present world. Ward regrets the fact that this model was not followed in his own country.

> Above all, this is strictly an Argentine museum. And it is in virtue of this distinctive characteristic that its collection holds special interest for the foreign visitor, in addition to its great importance for the Argentine people. With few exceptions, everything in the museum belongs to the great Argentine Republic,

whose natural history it illustrates and elucidates, from the remotest past to the present. It is to be lamented that our National Museum in Washington did not adopt such a plan, which would have given us a truly *national* museum. In his arrangement of the various halls, as in the classification of their materials, Sr. Francisco P. Moreno, the museum's founder and director, has adopted the theory of evolution as best describing the gradual emergence of the faunas of the extreme south of the American Continent. A visitor to the museum may thus begin by studying the forms of life particular to early geological epochs, then progress from fauna to fauna until he reaches the present age. (148)

As Ward notes, with some admiration, the central organizing principles of the museum are extinction and speciation, making this a unique place in its time for studying the rudiments of evolutionary theory. Its instructiveness is further reinforced by the richness of the successive fauna of South America, all represented in the vast paleontological collection, many of them completely unknown to Ward, himself hardly a neophyte; "it would never have occurred to me, who am perfectly familiar with all the great museums of the world, that I might here be confronted with so many entirely new forms" (148).

Within the peripheral context of Argentina, an evolutionary approach served to mark the importance of loss, the necessity of destruction as a precondition for the reconstruction of the modern nation, which thus came to symbolize, not life, but death. The present became a transitional moment in evolutionary time, the space in which we verify the passing of species. This is not a museum that expresses arrival at some superior destination, as in Europe, but rather the struggle to get there. As we have discussed, this meant that instead of representing the presence of superiority, the institution is concerned with the absence of present self. The obsession with measuring, statistics, and population studies reflected the concern with marking the passage of biological populations. Death and waste were now the signs of progress. There was the temporality of progress, linear and consecutive, governing those who persisted; and there was also the time of those who were disposable and would not persist. Indians, for example, were always represented as dead and invisible, even when they remained alive, in significant numbers, throughout the country.

In this connection, it is worth noting that in 1992 indigenous groups brought suit to reclaim the remains of Chief Incayal from the Museo de la Plata. Irina Podgorny has described the process whereby such groups, which had never actually gone extinct, sought to reclaim their ancestors.[7]

Some of these ancestors, while still alive, were invited by Moreno to take refuge in the Museo de la Plata itself. One wonders what they thought, gazing at the remains of their relatives in their glass cases. In the Museo de la Plata, extinct lineages are all presented as severed branches of a common national trunk, reconciling their discontinuity with the larger-scale continuity of Argentina itself. In such a place, what is past is also present, while the present becomes a multiplicity of temporal overlays. The rupture of linear time embodied in this representation of evolutionary process leaves Ward, like Ameghino in his vision, in a state of altered consciousness.

> Readers of this necessarily superficial review will gain, at best, a very pale notion of the paleontological treasures of this great museum. They will doubtless be surprised to hear me declare that, in none of the public or private museums in the United States today, nor in any of the museums in the capitals of Europe when I last visited them in 1885, is there so vast a collection of large, mounted fossils, of any order of mammals, as in the Museo de la Plata. So startled was I by how much it contained, that on my first visit I entertained the daydream that I had been permitted to sample the delicacies of some fantastic vision. Only after repeated visits was I able to persuade myself that they were all, in fact, real. (H. Ward 1890–1891, 149)

As Moreno doubtless intended, Ward perceives the museum's anthropological exhibits as the logical extension of the paleontological collection. In the five halls dedicated to Argentine anthropology and archeology, "Argentine man, both modern and prehistorical, is represented by no less than eighty mounted skeletons and a thousand skulls." The only comparable collections are those of the Peabody Museum in Cambridge, Massachusetts, "the Saint-Germain Museum, near Paris, and those of London and Copenhagen" (H. Ward 1890–1891, 149). In anthropology, as in paleontology, the museum's strength lies in the *variety* of the individual specimens it contains. This is in keeping with its Darwinian approach, on which variation is the cornerstone of evolutionary change.

The culture of extinction to which this chapter has been devoted had its heroes, and their heroism was of a peculiar brand. Theirs were not models to be emulated, nor were they heroic by virtue of any exemplary behavior. Far from paradigmatic, their status was iconic. In Moreno's accounts of his Patagonian expeditions, we find their story woven, together with the constituents of evolutionary science, into an iconography of the Patria (Molloy 2001, 55). Three decades later, during the centennial celebrations of

1910, Moreno's collection of human remains, stolen in his obsessive profanation of Indian cemeteries, was used to build a notion of vertical temporality, a progressive evolutionary time for the nation. "In this spirit, it is interesting to note that after the 1910 centenary the museum's anatomical collection is renamed 'Pateón de los Héroes Auctóctonos' [Pantheon of Autochtonous Heroes]. Today an Indian, tomorrow a national ancestor" (Molloy 2001, 54). Their extinction became grounds for both rejoicing, because it proved that success in the production of waste would make the country civilized, and for despair, because it proved that progress must always leave death in its wake. The constant threat of being caught up not in the vertical time of progress, but in the changeless, flattened, horizontal time of the disappeared—the African, the *criollo*, and the Indian—would dominate the culture of the late nineteenth century and the beginning of the twentieth, when the young nationalists and their contemporaries desperately yearned for a renewal of faith in the future.

This multiply voiced temporality that reflected at the same time presence and loss, optimism and pessimism, and life and death, expressed a melancholy of race that dominated the intellectual discourse of many peripheral regions following the Darwinian revolution. The work of Anne Anlin Cheng is helpful in understanding this process. In *The Melancholy of Race*, Cheng addresses the consequences of racism in the subordinated subject: "This is racial melancholia for the raced subject: the internalization of discipline and rejection—and the installation of a scripted contest of perception. The invisible man's racial radar, at once his perspicacity and his paranoia, is justified. For the invisible man is both a melancholic object and a melancholic subject, both the one lost and the one losing" (Cheng 2001, 17).

According to Cheng, the nature of the invisibility that emerges out of this system of exclusions is vital to any reading of the affected bodies and to understanding the melancholic incorporation "of the self as a loss" (17). To attempt to construct or project being is to confront the necessity of narrating loss and acknowledging the failure of totality. Within this void, the absent becomes the subject itself, nourished by its consumption of absence.

To paraphrase Cheng, in the melancholic landscape, loss becomes exclusion. This operation demands, first, that the melancholic subject reject loss *as loss* in order to preserve the fiction of possession, a rejection readily observed in turn-of-the-century authors confronting the destiny of the Afro-Argentine population and Indian population. Second, the melancholic subject must assure itself that its object will never return, so as not to endanger the cannibalistic cycle engendered by its absence. In this connection, Cheng's citation from Thomas Mann applies particularly well to the Argentine relationship with the lost body of the other: "What we call mourning for our dead is perhaps not so much grief at not being able to call them as it is grief at not being able to want to do so" (9).

CHAPTER FIVE

Sexual Selection and the Politics of Mating

Faced with anxieties provoked by the prospect of extinction, Argentine intellectuals operating in the post-Darwinian context had several options. Like Ameghino, Moreno, and Bunge, they could accept the darker consequences of evolution, acknowledge the likely interruption of their own biological posterity, and console themselves by vesting their hopes, instead, in the future of the nation. Or they could attempt to reconcile the basic tenets of evolution by natural selection with the Enlightenment optimism inherited from an earlier generation. Those taking the second approach were helped by the increasing importance of sexual selection within the Darwinian orthodoxy following the 1871 publication of *Descent of Man*. By clever reasoning, such figures as Domingo Sarmiento and Eduardo Holmberg attempted to show that nation and race might be understood as having emerged by sexual selection, and thus subject to cultural direction. They thus provided a potential avenue of escape from a hard biological determinism that dwelled on the inevitability of atavism and extinction. Sexual selection allowed beauty, empathy, and idealism to operate alongside struggle and competition, sanctioning the hope that the mitigating influence of the former might outweigh the implacable cruelty of the latter. For Sarmiento, sexual selection thus became the lynchpin of an attempted reconciliation between Darwinism and the Humboldtian ideal of science on which the members of his generation had been weaned.

In its April 1871 issue, the *Quarterly Journal of Science* published a review of Charles Darwin's *Descent of Man*, a work that, since the advance announcement of its publication "has for the last year and a half stimulated the curiosity and excited the most intense interest in the whole scientific world" ("Descent of Man" 1871, 248). The (unnamed) author of this review begins his discussion by noting, with a hint of disappointment, that the question of the descent of man occupies a relatively small portion of Darwin's book, with the rest given over to sexual selection. Still, he admits, the latter topic presents an interesting array of problems, to which Darwin has applied his considerable talents. Chief among these, the reviewer argues, is the problem of explaining the evolutionary emergence of an aesthetic sense among females, which in turn underwrites female mate choice. He asks, "Is not the mental development of the female, in fact, a harder problem to solve than the physical development of the male?" (253). In the debates that followed the publication of *Descent of Man* in Europe, the issue of female mate choice, and with it, what Haeckel had called "female selection," was among the most contentious. Curiously, however, it was far less prominent in Argentina. Of greatest importance here on the periphery was the fact that sexual selection reintroduced a causal role for the human will, with its aesthetic values, in directing the progress of the human species. The project of national repopulation could thus be undertaken with the knowledge that its outcome would not be determined by cruel, blind natural selection alone.

It is worth noting that comparable applications of the concept of sexual selection may be found throughout Spanish America in the late nineteenth and early twentieth centuries. Beauty became the crucial selective force in the improvement of the nation. As an example of how widespread this attitude was in the former Spanish colonies, consider a speech delivered in 1903 by the Peruvian doctor Wenceslao F. Molina.

> Lamarck, Darwin, and Haeckel assert that the species, and therefore races are subject to the law of evolution by triple selection, natural, sexual, and artificial, determined respectively by the struggle for life, sex, and man....
>
> Sexual selection, in essence merely a form of natural selection, modifies individuals by transmitting aesthetic qualities with powerful influence over the perfection and diversification of races. (1903, 8)

It is worth asking how a mechanism that Darwin had introduced to explain "a struggle between the males for the possession of the females" that resulted not in death "to the unsuccessful competitor, but few or no offspring," ended up so intimately associated with the transmission of beauty (C. Darwin 2006, *On the Origin of Species*, 506). We believe that, in this type of selection, those concerned with the consequences of natural selection, and especially with atavism and extinction, found their escape. This was, after all, a more benign form of evolution, more consonant with Enlightenment and Romantic conceptions of social progress.

Discounting female agency in mate choice, women were to be placed as decoys to attract the best men to become the fathers of the nation. Sexual selection also sustained the belief that human will and desire, as expressed in the aesthetic sense, played an important role in shaping the future, as well as underwriting normativity in general. Along with the importance of beauty, Friederich Nietzsche's influence on prevailing notions of human will and aspiration also played an important role in the development of this particular interpretation of sexual selection by the end of the nineteenth century (see Novoa and Levine 2009). We can see traces of this line of thought elsewhere in Spanish America. The Uruguayan José Rodó, author of *Ariel*, published in 1900, explains certain racial transformations by analogy to mating behavior in birds, as described by Darwin. And in 1925, in his *La raza cosmica* (The Cosmic Race), Mexican writer José Vasconcelos ascribes crucial importance to aesthetic selection in the prophesied emergence of the "cosmic race" (Vasconcelos 1997; see also Novoa 2009b).

In *Descent of Man*, Darwin offered a new explanation for the origins of racial difference, which could now be understood as a consequence of the application of aesthetic preferences in mate choice. The attraction of this explanation derived, at least in part, from the fact that sexual selection in the struggle to reproduce was perceived as considerably more benign than natural selection in the struggle for existence. It was also far more congruous with a pre-Darwinian conception of the evolution of human society, as striving toward beauty and harmony. Sexual selection was thus hugely appealing to Spanish American societies engaged in nation-building projects. Where natural selection left little room for the operation of human agency, sexual selection allowed, at least in principle, for the self-conscious realization of an ideal aesthetic type. Nations and races could thus be deliberately brought into being, molded into biological conformity with national values. While Darwinian evolution is ordinarily understood as undermining the reality, or at least the immutability, of species types, in Argentina, as elsewhere in Spanish America, sexual selection was understood as a tool for the *construction* of such types, in accordance with cultural preferences.

Nor was this deployment of sexual selection peculiar to this region. Edward Bellamy's *Looking Backward* (1888), the third-best-selling American book of the nineteenth century, pulled together "themes associated with Saint-Simonian socialism, Comte's Religion of Humanity, Marx's *Capital*, and Darwin's *Descent of Man*" (Olson 2007, 265). In Richard Olson's words, Bellamy saw a "trend toward increasing selflessness and love of all humankind, which he viewed as the consequence of Darwinian sexual selection" (265). Darwinism in general lent itself well to such eclecticism. As Darwin's critics, from William Whewell and John Stewart Mill to Hermann Burmeister had noted, Darwin's hypotheses were tied to his empirical evidence not by any of the standard methods of inductive argument, but by a collection of theory-constitutive analogies (see introduction). The open-endedness of these analogies allowed them to be sustained by a great variety of evidence—witness Darwin's own eclecticism in his selection of sources—but it also, inevitably, meant that both source and target of analogy were tied to other, equally open-ended concepts within the cultural webs that gave them meaning. To those who study the history of ideas in Europe and North America, it is easy to take the rest of the web for granted.

The study of peripheral science is thus of value in part because, by illuminating the fate of theory-constitutive analogies in cultural contexts in which other concepts in the web are less familiar, it reveals the very existence of such webs, and the operation of analogies within them. In the work of the Argentine thinkers discussed in this chapter, as in Bellamy's *Looking Backward*, sexual selection operates as a kinder, gentler mode of evolutionary change. But unlike Bellamy, these thinkers saw themselves as belonging to populations threatened with extinction. Their need for a tonic for the brutality of natural selection was consequently much more urgent, and their evolutionary thought had to respond to this urgency. Directed toward social policy, it offered the possibility of salvaging national continuity and directing the course of population change by controlling who would be allowed to reproduce. The Argentine counterpart to North American and European eugenic policies was a policy of selective mating, in which the politics of mate choice were seen to be essential to the fate of the modern nation.

Sexual Selection in Darwin

Darwin coins the phrase "sexual selection" in chapter 4 of the 1859 edition of *Origin*, but his discussion of it is relatively cursory. It serves as a potential explanation for the persistence of traits whose presence is difficult to account for by natural selection alone. Among certain birds, for example, there is a kind of selection predicated not on the struggle for existence, "but on a struggle between the males for the possession of the females; the result is not death to the unsuccessful competitor, but few or no offspring" (C. Darwin 2006, 506). The force of such selection is thus "less rigorous than natural selection. Generally the most vigorous males, those which are best fitted for their places in nature, will leave most progeny" (506). It is effective to the extent that "individual males have had, in successive generations, some slight advantage over other males, in their weapons, means of defence, or charms; and have transmitted these advantages to their male offspring" (507).

The hints in *Origin* clearly demanded further elucidation. In a work that would become very influential in Spanish America, Ernst Haeckel offered a substantive discussion of this type of selection. His *Generelle Morphologie der Organismen* (General Morphology of Organisms; 1866) was highly praised by Darwin in his letters to the German naturalist and in his introduction to *Descent of Man*. Generalizing the account of sexual selection in *Origin*, Haeckel notes that just as males may struggle for possession of females, so may females "struggle for possession of males," a process Haeckel terms "female selection" (Haeckel 1866, 244). Spurred by Haeckel, among other supporters and critics, Darwin bent his own efforts to developing a fuller account of sexual selection. These bore fruit with the 1871 publication of *Descent of Man*. In his preface to the second edition of 1874, Darwin responds to the criticism that when he had "found that many details of structure in man could not be explained through natural selection," he had "invented sexual selection." He reminds his readers that the concept had been introduced in *Origin*, in which he had also "stated that it was applicable to man" (C. Darwin 1874, vi), sparking a significant debate among both supporters and opponents of his theory of evolution. We will confine our discussion to the specific issue of the applicability of sexual selection to humans, and to the emergence of race in particular.

The origins of the diverse races of humanity had been a subject of heated debate since well before the publication of *Origin*, along with the closely related question of the original unity or plurality of the human species. The conclusion that many readers of the *Origin* drew from the application of its account of natural selection to humanity was that racial differences must have originated in some collection of heritable advantages in the struggle for existence. Both the struggle itself and the various advantages that might be brought to bear in it were, it was argued, still in evidence in the nineteenth century; witness the success of the European races in exercising dominion over the rest of humanity. Unlike *Origin*, however, chapter 7 of *Descent of Man* addresses the "Races of Man" directly, attributing their differences to the operation not of natural selection, but of sexual selection.

Darwin begins this chapter with the disclaimer that, unlike many others who have treated this subject, it is not his "intention . . . to describe the several so-called races of men; but to inquire what is the value of the differences between them under a classificatory point of view, and how they have originated" (C. Darwin 2006, 899). With regard to their description, he is quick to point out, there is no general consensus. In chapter 2 of *Origin*, he had rested the argument of the entire book on the fact that naturalists cannot, in general, agree on when two organisms belong to different species, rather than merely to different varieties of the same species. Now he observes:

> Man has been studied more carefully than any other organic being, and yet there is the greatest possible diversity amongst capable judges whether he should be classed as a single species or race, or as two (Virey), as three (Jacquinot), as four (Kant), five (Blumenbach), six (Buffon), seven (Hunter), eight (Agassiz), eleven (Pickering), fifteen (Bory St. Vincent), sixteen (Desmoulins); twenty-two (Morton), sixty (Crawford), or as sixty-three, according to Burke. This diversity of judgement does not prove that races ought not to be ranked as species, but it shews that they graduate into each other, and that it is hardly possible to discover clear distinctive characters between them. (905)

In both *Origin* and *Descent of Man*, such gradation is cited as clear evidence of common descent. In opposition to those who had argued, with Agassiz and Burmeister, that the differences between human races were so marked as to force us to assume a plurality of species, each with its separate origin, Darwin insists on original unity. But once species unity is presupposed, the tremendous diversity of the (single) species becomes apparent. There may be no "clear distinctive characters" for distinguishing races, but there *is* a tremendous range of heritable difference, including differences in skin and hair color commonly cited as racial markers. "We are therefore led," says Darwin,

> to inquire whether slight individual differences, *to which man is eminently liable*, may not have been preserved and augmented during a long series of generations through natural selection. But here we are at once met by the objection

that beneficial variations alone can be thus preserved; and as far as we are enabled to judge ... not one of the external differences between the races of men are of any direct or special service to him. ... In this respect man resembles those forms, called by naturalists protean or polymorphic, which have remained extremely variable, owing, as it seems, to their variations being of an indifferent nature, and consequently to their having escaped the action of natural selection. (917; emphasis added)

The human species remains "protean" because most of the individual differences among its members convey no advantage in natural selection. To the extent to which this species has varieties or races, the differences between them cannot, Darwin argues, be traced to the different selection pressures operating in the wide range of environments they inhabit. "We have thus far been baffled in all our attempts to account for the differences between the races of man; but there remains one important agency, namely Sexual Selection, which appears to have acted as powerfully on man, as on many other animals ... the differences between the races of man, as in colour, hairyness, form of features &c. are of the nature which it might have been expected would have been acted on by sexual selection" (917–18). The precise manner in which sexual selection has operated in human evolution remains a matter of speculation, Darwin admits. Many of his conjectures are motivated by analogies to animal models, especially birds. Significantly, this includes his conjecture regarding the evolution of the human aesthetic sense. "On the whole, birds appear to be the most aesthetic of all animals, excepting of course man, and they have nearly the same taste for the beautiful as we have. This is shewn by our enjoyment of the singing of birds, and by our women, both civilized and savage, decking their heads with borrowed plumes, and using gems which are hardly more brilliantly coloured than the naked skin and wattles of certain birds" (1044).

It would be hard to overstate the importance, for present purposes, of the role Darwin ascribes to beauty in sexual selection among humans. For the sense of beauty is not merely variable, it is subject to cultural determination. Beauty is the wedge whereby culture enters into biological evolution.

> The taste for the beautiful, at least as far as female beauty is concerned, is not of a special nature in the human mind; for it differs widely in the different races of man, and is not quite the same even in the different nations of the same race. ... Obviously no animal would be capable of admiring such scenes as the

heavens at night, a beautiful landscape, or refined music; but such high tastes are acquired through culture, and depend on complex associations; they are not enjoyed by barbarians or by uneducated persons.

Many of the faculties, which have been of inestimable service to man for his progressive advancement, such as the powers of the imagination, wonder, curiosity, an undefined sense of beauty, a tendency to imitation, and the love of excitement and novelty, could hardly fail to lead to capricious changes of customs and fashion. (814)

In Argentina, sexual selection, operating through the evolved aesthetic sense, sanctioned the revival of the pre-Darwinian culture of civilization, inspired by the revival of *Naturalphilosophie* and romanticism in Hackel's evolutionary work. As we have noted, Ernst Haeckel revived the scientific concern with aesthetics. As Gregory Moore puts it, Haeckel used "his concept of the Kunsttrieb to play down the importance which Darwin attached to natural selection, and instead portrayed evolution as a linear progression towards ever higher, more beautiful organic forms" (G. Moore 2002, 103). Moreover, Haeckel's use of sexual selection enhanced a gender categorization that paired men with intelligence and women with beauty, as Robert J. Richards explains. "Prior to the discussion of human sexual selection that would occupy Darwin's *Descent of Man* [1871], Haeckel maintained that within the advanced races, females would select men of higher mental caliber, thus continuously increasing the brainpower of the species. In like fashion, active male choice would enhance female beauty" (Richards 2008, 158). For Argentine intellectuals, these consequences, taken together, meant that they could be Darwinians while at the same time preserving the idealist thought so closely tied to their process of nation building. As Ernst Mayr has pointed out, in evolutionary theory, and especially in the Modern Synthesis, it is difficult to define sexual selection as a separate process distinct from natural selection, as Darwin appears to have wanted to do—over the objections of Alfred Russell Wallace and others (Mayr 1985, 596). Regardless of the ultimate verdict of the scientific community regarding the biological respectability of sexual selection, the vast impact of this concept in the social sciences, as evident in Argentina and elsewhere, is undeniable. In Argentina, it acquired a political meaning closely tied to the project of repopulating the country with a new race, instantiating the racial type appropriate to the national aesthetic sensibility. Recast in the context of this project, the reach of Darwin's analogies ultimately extended into literature and popular culture.

The Analogies of Sexual Selection in Argentina: Sarmiento and Holmberg

As we discussed in the previous chapter, interest in natural selection—and in its darker side, extinction—was nearly universal among Argentina's naturalists and anthropologists. Because sexual selection was understood to have immediate consequences for the understanding of race, courtship, and mating, its constitutive analogies had even broader appeal, extending to the general literate public. The penetration of this concept into popular culture in the 1880s coincided with increased public attention to issues surrounding the sexuality of the youth, along with the social roles appropriate to the two sexes (Novoa 2008). It was believed at the time that in strictly *natural* selection, culture, including aesthetic culture, would have little role to play. Now, though, it seemed possible that *beauty*, the chief weapon in reproductive competition and one subject to cultural manipulation, might prove a more effective tool for the creation of a new national race.

Significantly, the social and political importance of mating practices had been argued for three decades earlier, in one of the foundational texts of the Argentine Republic. Juan B. Alberdi's *Bases y puntos de partida de la organización política de la República Argentina* (Foundations and Points of Departure for the Political Organization of the Argentine Republic), written shortly after the defeat of Rosas in 1852, laid much of the groundwork for the constitution of 1853. Like other members of the Generation of 1837, Alberdi had set out from a conception of civilization as a cultural order conducive to progress. By this time, however, his views had shifted toward a biological determinism, anticipating some aspects of the Darwinian revolution (see chap. 3).

In Alberdi's view, acquiring a modern population whose men displayed the necessary aptitudes was essential to establishing a modern nation. Such men were to be found in Europe. Alberdi recognized that they would need considerable inducement to relocate to a remote, primitive corner of the globe. Fortunately, Argentina's extraordinarily beautiful women could supply the necessary incentive. Lured by their charms, immigrant men would contribute their own desirable traits—their love of freedom and industry, in short supply among Argentine men, to the future Argentine population. Any Argentine men reluctant to cede the most beautiful women to foreigners ought to be consoled by the fact that, under Alberdi's plan, the Argentine racial type, which these same women embodied so delightfully, would survive.

We must exchange our people, unsuited to liberty, for others who are better suited, but without abdicating our original racial type, let alone our mastery of the country; we must supplant our present Argentine family in favor of another, equally Argentine, but more capable of freedom, wealth, and progress. Are we, then, to seek conquerors more illustrious than the Spaniards? To the contrary, we must conquer, rather than be conquered. South America possesses an army for this purpose: the charms its beautiful, admirable women inherited from their Andalusian foremothers, further improved under the splendid skies of the New World. Remove the immoral impediments that have sterilized the power of the American fair sex, and you will achieve the transformation of our race, without sacrificing either language or primitive racial type.

This change, gradual but profound, this racial transformation, must be the task of truly regenerative, progressive [South American] constitutions. In the American interest, they must initiate it and see it through, rather than leaving matters to the spontaneous action of a system of things that tends, instead, toward the gradual erosion of the ascendency of the Spanish type in America. (Alberdi 1856, 139)

In the constitution enacted the following year, immigration is enshrined as the foundation for establishing modern institutions. Immigrants presumed to be civilized are indirectly granted reproductive privileges. Argentine women are thus conceived as the receptacles for their desirable traits, while also being charged with the responsibility of preserving the Spanish racial type.

Alberdi, incidentally, was not alone in his assessment of the potential of South American women as an army of conquest. Nineteenth-century travel literature offers other examples of avid descriptions, with some of the greatest accolades reserved for the women of Buenos Aires. In 1820 the American writer Henry Marie Brackenridge reports that while making the rounds of the city's churches, he found his attention drawn more "by the crowds of beautiful women, going and coming to the churches, and the graceful elegance of their carriage. They walk more elegantly than any women I ever saw" (Brackenridge 1820, 260). Ever alert to the local environment, Woodbine Parish observes that some of the women of Buenos Aires "are very beautiful, and their polite and obliging manners, especially to strangers, render them doubly attractive." Anticipating Alberdi's strategy, Parish notes that his fellow Englishmen "have formed many matrimonial connexions with them, which has contributed, no doubt, to the good feeling with which they are so generally regarded by the natives" (Parish 1839, 32).

Sarmiento's *Civilization and Barbarism*, written in 1845, enjoyed broad influence throughout Spanish America. Translated into English in 1868 by Mary Mann, wife of Horace, it contains the same rhetoric about female beauty typical of travel narratives about Argentina, which Sarmiento read religiously. The orientalization of the local female population is a function of the association of Argentina with other regions encompassed within the imperial gaze, a perspective Sarmiento had internalized. "Perhaps one might believe this description to be taken from the 'Thousand and One Nights,' or other Eastern fairy tale; but I cannot half describe the voluptuous beauty of these damsels, daughters of the tropics, as they recline for their siesta beneath the shade of the myrtles and laurels, enjoying such odors as would bring asphyxia upon one unaccustomed to the atmosphere" (Sarmiento 1868, 205).

This same perspective will continue in the literature of the next century. A 1910 book on Argentina by William Alfred Hirst notes that "Argentine women have a reputation for beauty," now a commonplace assessment (Hirst 1910, 153). In 1918 Clayton Sedgwick Cooper offers, "No article on this city [Buenos Aires] would be complete without mention of the Argentine women; and it must be said, furthermore, that the women of Buenos Aires, both by their beauty and feminine charm, live up to the artistic standard of the Capital's home and general magnificence" (Cooper 1918, 257). But by this time such emphasis is often related with the idea of racial types. In a book on Argentina and Uruguay written by Gordon Ross, the former financial editor of the Buenos Aires English-language daily *The Standard*, the same stereotype appears, representing women as one of the country's assets. "Argentine women? This is a subject on which one is not only tempted but almost forced to confine oneself to the usual platitudes concerning beauty of the Spanish type: large-eyed and opulent and at its apogee during the decade between 15 and 25 years of age" (Ross 1916, 55).

By 1900 female beauty is regularly extolled as one of the modern aspects of Argentine culture, an essential proof of the success of the country's modernization process. This understanding of beauty is closely related with prevailing readings of Darwinism. Following the publication of *Descent of Man* in 1871, the perceived importance of female beauty as a weapon for the conquest of resources needed to civilize the country rose still further. On May 19, 1882, exactly one month after Darwin's death, the Buenos Aires Círculo Médico (Medical Circle) sponsored a hastily organized Darwin tribute, including lectures by two influential thinkers,

Domingo Sarmiento and Eduardo Holmberg. Both pieces would be widely circulated, and in both, sexual selection plays a prominent role. We will consider each in turn.

Sarmiento's lecture is of particular interest for the unique blend of pre-Darwinian and Spencerian thought with which he explains central tenets of Darwinism. Of all of the self-professed Darwinians, Domingo Sarmiento enjoyed by far the greatest political influence. Born in San Juan in 1811, he left Argentina for Chile in his youth, forced into exile by his opposition to Rosas. On his return to Argentina following Rosas's ouster, Sarmiento became a leading intellectual and political figure of his generation. He held various high posts, culminating in 1866 with his appointment as ambassador to the United States, and his term as president of the Republic from 1868 to 1874. While in the United States, he established relationships with some of the leading intellectuals of the day and participated in debates on evolution. Following the conclusion of his term as president, he remained an elder statesman of considerable political and cultural prominence until his death in 1888.

Sarmiento's eulogy for Darwin thus dates from a late period in the author's career, and not an easy one, as he found himself increasingly depressed and convinced of his personal failure. It is a perplexing text, and understanding it requires that the complexities of Sarmiento's relationship with Darwinism be taken into account. In it he argues, against such naturalists as Ameghino and Moreno, that Darwinism represents a continuation of older schools of thought and not a break with them. This position earned him the derision of proponents and opponents of Darwinism alike. The former chided Sarmiento for his lack of scientific rigor and for his seeming ignorance of the radical novelty of the principles on which the new theory rested. The latter, and most especially Catholic conservatives like José Manuel de Estrada, took him to task for affirming that humans had descended from apes. As we have noted, Sarmiento's need to maintain continuity with Enlightenment ideals also runs through his last major work, *Conflicto y armonias de razas en América* (1883), the subject of merciless critique in the Buenos Aires press. In his response, Sarmiento would call the book, "the last call of the reasoning, the principles and the tradition of May [the May revolution of 1810], which in turn was only the crest of a wave that has roiled the seas since 1776, traveling not from East to West, but from North to South, clearing and guiding us along the new path that opened to humanity with the advent of American institutions" (Sarmiento 1900g, 326).

Sarmiento campaigned actively to impose his own interpretation of evolution, both in Argentina and abroad. In an April 9, 1883, letter to Francisco Moreno, thanking him for coming to the defense of *Conflicto y armonías de razas* after its scathing review in *The Standard*, Sarmiento admits that, strictly speaking, his view of evolution is Spencerian, ceding Darwinism proper to Moreno and Ameghino. Despite his public defense of Darwinism, including the Círculo Médico lecture, in private he allowed that he had no right to call himself a Darwinian. Instead, "I get along with Spencer. We follow the same path" (Sarmiento 1900g, 322). Such admissions indicate that, notwithstanding the accusations of ignorance or confusion, Sarmiento clearly understood what was revolutionary about Darwinism. If not quite openly, he nonetheless self-consciously rejected the problematic tenets and presuppositions of this theory, including its hard materialism, because they demanded a traumatic break with the civilizing intellectual tradition he had spent his life defending.

But *qua* defender of civilization and science, Sarmiento had had no choice but to champion the *new* science. In this capacity, even Moreno charged him with neglecting its most important principles. Such charges missed the subtlety of Sarmiento's positioning vis-à-vis evolutionary theory. In the aforementioned letter to Moreno, Sarmiento acknowledges the national scope of the political controversy surrounding both the theory of common descent and Sarmiento's own association with it. He notes that in Córdoba, for example, a scandal had arisen because "Sarmiento says we're the sons of apes," while in Salta, "one Representative Ortiz spent a half-hour denouncing Sarmiento's insult in calling the Governors mulattos" (Sarmiento 1900g, 323). Such polemics were a response, in part, to Sarmiento's public lecture on Darwin and to his wide dissemination of the text of that lecture among the educated classes. In a letter to José Posse of May 18, 1882 (the day before the Círculo médico event), he promises to send a copy of his "lecture on Darwin, which is quite good" (Sarmiento 1961, 140). The letter continues with an analysis of the uses and abuses of Darwinism in the hands of various politicians.

Another letter to Posse five days later promises an entire "Darwin package, *to be sold* in Tucumán, Salta, and Jujuy" (Sarmiento 1961, 141; emphasis in original). The lecture on Darwin is to be distributed as broadly as possible, in persecution of Sarmiento's rivalry with erstwhile ally Nicolás Avellaneda, who had recently reviled both Sarmiento and Darwin in his efforts to curry favor with the Church. The Darwinian counterpropaganda must "circulate in the North. We have greatly aroused San

Luis and San Juan, while in Buenos Aires they sound the fanfares of enthusiasm. The enemy press has risen to the challenge!" (141).

The name of Darwin, it seems, was an appropriate one for Sarmiento to rally his supporters around. While prepared to confess his preference for Spencer over Darwin in a private letter to Moreno, he never mentions the former's name in what was, after all, a eulogy for the latter. Still, in lecture and pamphlet, Sarmiento professes an evolutionism of broadly Spencerian stripe: "I, gentlemen, believe in evolution as a spiritual process, because I find I must rest on a principle both harmonious and beautiful, setting aside all doubt, which torments the soul" (Sarmiento 1928, "Conferencia sobre Darwin"). Spencer's brand of evolution was much more congenial than Darwin's. His system made explicit many intuitions given voice in Sarmiento's own work as far back as *Facundo*, as it revived some of the pre-Darwinian evolutionary ideas that had influenced Sarmiento in his youth. A teleological evolutionary framework was compatible with the progressive perfection of society, the "spiritual process" to which Sarmiento refers. To a thinker preoccupied with social progress, the Darwin of *Origin* remained deeply problematic. To be sure, that Darwin had challenged the fixity of species, varieties, and, by extension, races; he had undermined species essentialism; and he had established the likely emergence of new species, varieties, and races in the future. But because evolution by natural selection always proceeded by variation and selective retention, lineages tended to *diverge*. On Darwinian grounds, a *converging* future, marked by increasing homogeneity and unity of the sort long envisioned by Sarmiento thus becomes extremely unlikely.

As in the Círculo Médico euology, in *Conflicto y armonías* Sarmiento weaves Darwinism into a grand totalizing theory, something Darwin himself would never have attempted. Darwin discovered a universal mechanism—natural selection—and explained how, depending on the particular circumstances in which it operated, it might give rise to very divergent consequences. What Sarmiento wanted was a universal principle that would always produce the same consequences over a given, determinate span of time: a *principle of historical necessity*. He views the absence, in Darwin, of any historical narrative, as a defect to be remedied. In this book he undertakes the correction, helping himself to Darwin, but also to Antonio Snider-Pelegrini, Florentino Ameghino, and Schliemann's Troy, weaving them together into a kind of mythical Atlantis story, explaining the similarities between Africans, Americans, and Asians. Thus is the grand universal narrative restored.

If his attempt is diffuse and chaotic, this is not due to ignorance or inattention to his sources—quite the contrary. Sarmiento, we claim, understood full well the magnitude of the Darwinian revolution and strove to supplement or correct it by falling back on the dearly held views of the past. Sarmiento's own agenda is only partially to blame for the confusion. Another part must rest with *Origin* itself. Arguing so persuasively in support of a theory that rested, ultimately, on analogies to variation and selection under domestication and that lacked any account of the all-important mechanisms of inheritance was no easy task; even Darwin's best efforts left critics and supporters alike concerned for those mechanisms (Vorzimmer 1963). Spencer's neo-Lamarckianism is extremely useful in this context, as it revives soft inheritance, an attractive concept for those interested in constructing the race of the future.

For a closet Spencerian saddled with a political obligation to defend Darwin, *Descent of Man* had thus appeared as a godsend. Sarmiento latches onto Darwin's understanding of sexual selection as a way of recovering a strong evolutionary role for soft inheritance within a framework driven by natural selection. Darwin's discussion of sexual selection allows Sarmiento to claim that Darwin had, after all, recognized the importance of beauty and harmony in the evolution of civilization. And if the deployment of aesthetic criteria in mate choice has evolutionary force, it strongly suggests the survival value of a morality based on taste. Sexual selection thus becomes the lynchpin both of Sarmiento's restoration of the teleology of civilization and of his claim to represent Darwin and Darwinism. The pursuit of Argentina's evolutionary hopes becomes analogous to a great dance, in which the national population, like a flock of birds, employs its sensitivity to beauty as a mode of selection. The improvement of this population thus becomes, once more, a process in which individual free will has a role to play, one that trumps blind natural selection. Like Alberdi in the 1850s, Sarmiento in the 1880s pins his hopes for the future on reproductive choices predicated on feminine beauty.

> As anyone who has seen the spectacle of an aristocratic ball can attest, when young women move to the cadence of the music, they instinctively confirm Darwin's theory. They improve and beautify the race by means of all the attractions and seductions, all the colors, shapes, and adornments the fine arts can offer. A spray of flowers, or feathers, carelessly draped to one side of the head, causes her to lose her balance and posture, lifting her face in compensation, and the undulation and inclination of her swanlike neck display vivacity and

intelligence. The peacock's tail has instructed queens and princesses in their conceptions of majesty, and when a girl spends her time training her posture before the mirror, eventually she will become beautiful, or if not her, then her daughters, until the poise and elegance brought to America by the Andalusian ladies, and still preserved in our older families, become hereditary. Darwinism, pure Darwinism, is nothing more than what the vulgar, *antiscientific* mouths call coquetry, from "cock," or rooster, referring to the bad habit of primping and fluffing one's plumage. (Sarmiento 1928, 165)

Significantly, despite all the attention he pays to the evolution and cultivation of feminine wiles, a key aspect of Darwin's understanding of sexual selection is conspicuously absent from Sarmiento's account. Unlike Darwin, he does not see, or does not wish to present, *male* adornments as *weapons* in a fierce competitive struggle for reproductive privilege, a struggle in which there are winners and losers. Nor does he consider the fact that, like favorable traits in the struggle for existence, such adornments must initially arise as chance variations. Like that of evolution by natural selection, the outcome of evolution by sexual selection is ultimately *contingent*; as Darwin saw it, it does *not* underwrite a principle of historical necessity. While superficially Darwinian, Sarmiento's attachment to beauty is really profoundly romantic. As the dance analogy suggests, far from weapons, the traits of a beautiful woman are simply salient features of a world moved by deep, underlying harmony.

Sexual selection played a pivotal role in the eclectic view of biological, cultural, and historical progress Sarmiento developed late in his life. This approach met with mixed reviews. In the 1882 *Anuario bibliográfico de la República Argentina* (Bibliographic Yearbook of the Argentine Republic), the author of a capsule critique of the published version of Sarmiento's Círculo Médico lecture calls it "uneven, like everything the author has produced in this period of his life. The lecture contains brilliant and profoundly insightful passages, side-by-side with others of real decadence, either obscure or frankly crude" (Darwin en una conferencia 1883, 214–15). Sarmiento's eclecticism, unsuccessful though it may have been, was a deliberate, self-conscious effort, one that reflected the complex intellectual trajectory of an author who strove, to the last, to remain faithful to the Enlightenment ideals of his youth. As such, it also reflected the intellectual trajectory of nineteenth-century Argentina in its coming to terms with successive Darwinian principles. Sarmiento is prepared to project an analogous developmental history onto Darwin himself, hypothesizing that the

English naturalist, dissatisfied with the limitations of an account of evolution that attributed "the variations of organic forms to the natural selection of the most vigorous types, and those most adapted to the struggle for existence in their respective environments," had therefore proceeded to identify "an even more significant cause, the sympathetic aspiration toward beauty, by which so many animals have clothed themselves in such exquisite forms, adorned with inimitable elegance and luxury" (Sarmiento 1928, 166).[1] This attribution is of interest less for Sarmiento's insight into Darwin's motives—Darwin's own path to *Descent of Man* was doubtless a different one—but for what it reveals about Sarmiento's intellectual needs and those of his generation.

The critical review of Sarmiento's Círculo Médico lecture in the *Anuario bibliográfico* appears next to a much more favorable account of Eduardo Holmberg's contribution to the memorial event, which had also appeared in pamphlet form (Holmberg 1882). Where Sarmiento's efforts are described as, at best, "uneven," Holmberg's is "methodical, clear, and correct. . . . This piece is worthy of the intelligence of the young Argentine naturalist" ("Cárlos Roberto Darwin, por Eduardo Ladislao Holmberg," 214). Holmberg was born in Buenos Aires in 1852, grandson of an Austrian baron who had served with José de San Martin and Carlos María de Alvear during the Napoleonic Wars and subsequently joined their revolutionary cause. His penchant for combining scientific and literary pursuits became apparent when he was still quite young, with the publication in 1872 of his account of a trip through Patagonia, *Viajes por la Patagonia* (Travels through Patagonia). In 1877 he traveled throughout the country, publishing several articles on his zoological and botanical discoveries in various scientific journals. His botany textbook remained a standard of the Argentine secondary school curriculum for many years.

Holmberg graduated from the School of Medicine of the Universidad de Buenos Aires in 1880, though he never practiced, preferring scientific and literary pursuits.[2] One of the most well-known Darwinians of his generation, in his zoological and botanical work he regularly defended Darwinism from its detractors, who had rallied around Burmeister. He was able to promulgate Darwinism *ex cathedra* from his position as the first professor of natural history in the Facultad de Ciencias Exactas y Naturales of the Universidad de Buenos Aires. He founded *El Naturalista Argentino* and was a frequent contributor to the *Revista Americana de Historia Natural*. He was also the founding director of the Buenos Aires Zoological

Gardens, departing in 1904. At his death in 1937, he was among the most celebrated figures in Argentine science.

Like most of the figures considered in this book, Holmberg was a member of the Generation of 1880, a generation that struggled to reconcile the culture of civilization, as promulgated and defended by Sarmiento's Generation of 1837, with the realities of their own situation. Holmberg's work constituted an important contribution to this struggle and remains a central piece of evidence regarding the reception of Darwinism in Argentina and its importance in the cultural, social, and political life of the nation. Unlike Florentino Ameghino, stigmatized for his humble, immigrant origins, but like Francisco Moreno, Holmberg was a scion of one the most aristocratic families of *porteño* society. As such, much of his work must be understood as contributing to an intellectual project of his class, that of exerting intellectual leadership to transform a primitive country into a truly civilized country.

In the first part of his eulogy, Holmberg devotes a great deal of attention to explaining Darwin's analogy (in *Origin*) to domestic selection. It quickly becomes apparent that Holmberg, unlike Sarmiento, is a Darwinian of fundamentally Haeckelian stripe (see chap. 3 for further discussion of Haeckel's importance in Latin America). Like the neo-Lamarckianism so apparent in Sarmiento, Haeckel's physiological holism constituted another attempt to reconcile the ideals of a prior generation—in his case, those of the Humboldtian *Naturphilosophen*—with the Darwinian revolution. As David Paul Cook has put it, Haeckel "accepted the role of struggle in selection theory, a precondition to progress, but he used physiological parallels to underline the nationalistic principles of integration and mutual dependence in social evolution" (Cook 1994, 239).

Like Haeckel, Holmberg acknowledged the consequences of Darwinian analogies to a degree well beyond what Sarmiento could accept, endorsing a model of biological change grounded in breeding, struggle, and Malthusian logic and devoid of providential intentionality. But he also adopted Haeckel's Humboldtian synthesis, according to which there remained a special place in the theory for humanity and thus for human sentiment, reason, and will. As Peter Bowler has noted, the resulting model was profoundly teleological and explains why "some commentators saw Haeckel's evolutionism as little more than a superficially materialist gloss on the old idealism or transcendentalism of German *Naturphilosophie*" (Bowler 1996, 75).

Haeckel and his followers "found it difficult to break away from the inherently progressionist image of a ladder of developmental stages" (Bowler 1996, 75). This difficulty was not a matter of falling, unwittingly, into old habits of thought; the old habits were acknowledged, and dearly held, and much effort went into preserving them in the face of their incompatibility with Darwinism. In Argentina, as we saw with Sarmiento, the difficulty was compounded by the fact that science itself had been enshrined as essential to national progress, but the theory of evolution by natural selection threatened to undermine the very future of the nation. It might be possible, with effort, to interpret the phylogenies of certain lineages as progressive—the fossil records of many phyla and classes were routinely read as exhibiting trends toward increasing complexity—but so long as evolution proceeded by random variation and selective retention, such progress must always be *contingent*. As discussed above, the appeal of Spencer to someone like Sarmiento was plain, for Spencer accommodated Darwinian principles into a philosophical system in which progress became *necessary*. Haeckel appealed to a younger generation, but for analogous reasons.

The following example may demonstrate just how thoroughly Haeckel had penetrated Holmberg's understanding not only of evolution, but also of Darwin himself. In his Círculo Médico lecture, Holmberg elects to illustrate chapter 3 of *Origin*, "The Struggle for Existence," by means of an account of the recent success of an invasive thistle species in supplanting native thistles across the Pampas. This case is of particular interest to his audience, because the outcome of the competition among thistles has immediate consequences for the quality of pasturage and thus for the all-important cattle industry. As is well known, such ecological interconnections feature prominently in chapter 3 of *Origin*. Darwin argues that the success of field mice in their struggle to avoid feline predation must affect the distribution of red clover, for mice eat bumblebee combs and only the bumblebee fertilizes red clover flowers (see C. Darwin 2006, 496–97). After explaining the case of the invasive thistle, Holmberg refers back to the red clover example. But, significantly, the published account he cites is not Darwin's—with which he was certainly familiar—but Haeckel's, from the French translation of his *Natürliche Schöpfungsgeschichte* (Natural History of Creation) (Holmberg 1882, chap. 3). For his part, Haeckel makes it perfectly clear that his own knowledge of the case is entirely secondhand, deriving from chapter 3 of *Origin*. We can see in this example how the reception of Darwin's ideas was mediated by other scientists who better addressed local political and cultural concerns.

After explaining natural selection, Holmberg's talk proceeds, like Sarmiento's, to discuss the difference between natural selection and sexual selection, attributing observed and potential changes in the Argentine population to the operation of the latter. He saw, quite clearly, that sexual selection left human evolution, and with it, the fate of the races of humanity, subject to the vagaries of culture.

> Let us turn to an example, which though somewhat Spartan, may nonetheless serve as an illustration.
> Suppose that by some whim men come to prefer blue-eyed women, and mercilessly reject those with brown or black eyes.
> After several generations, all women have blue eyes.
> Now suppose women come to prefer blond mustaches in men. Again, after several generations, all men will have blond mustaches, if they have mustaches at all.
> Darwin's claims in the *Origin of Species* were not quite so bold. But later, with the publication of *Descent of Man*, he became much more daring. (Holmberg 1882, 130)

Attentive to his audience, Holmberg had offered a local illustration of the struggle for existence. He allows local color to bring sexual selection to life, too, but with an added *frisson*, as Holmberg's example draws attention to salient features of the audience itself. For this purpose, he cites a world-renowned authority; "The delicious pages devoted to the subject of *porteña* women by [Paolo] Mantegazza, applicable as they are to Argentine women in general, serve to motivate further inquiry into a possible application of the principle of sexual selection" (130). Mantegazza (1831–1910) was an Italian physician (practicing in Argentina from 1854 to 1858, where he married his first wife), Darwinian anthropologist, and pioneering psychopharmacologist whose published accounts of his travels in South America enjoyed broad circulation (Mantegazza 1876). As Nicoletta Pireddu has noted in her introduction to a recent translation of Mantegazza's *Physiology of Love*, "His professional versatility and volcanic personality won him the recognition of leading European intellectuals, among them Charles Darwin, of whom Mantegazza became an enthusiastic follower, and by whom he was rewarded in his turn by citations in . . . *The Descent of Man*" and other works (Mantegazza 2008, 8).[3] Mantegazza would become known as something of an authority on sexual selection in humans and is repeatedly cited as such in the classic *Psychology of*

Sex, by Havelock Ellis, precisely because of Ellis's interest in this subject (Ellis 1906; see, e.g., 82).

Holmberg quotes at length from a translation of a chapter of Mantegazza's in the popular literary magazine *La ondina del Plata*. The following excerpt will suffice to convey the flavor of Mantegazza's observations and of the interest they held for Holmberg and his audience.

> Andalusian blood runs through the *porteña*'s veins. She possesses every seductive quality grace or ingenuity can confer, and though she hides them well, she is gifted with all of the physical perfections the Arabs expect of the feminine sex. Her perfume is that of a hothouse flower, more guessed at than actually perceived, and she wields with skill the myriad intangible artifices of that most difficult and dangerous of arts, that of awakening desire, of being and not-being, of transporting men to circles of Paradise Dante never trod, where in their happiness and impatience, they are far removed from life's tedium. Yet at the same time, beneath her civilized, sophisticated veneer lurks a woman, Eve's daughter, whose robust figure and vigorous flesh bespeak an excellent mother and no less excellent wife. Her artful trappings remain inferior to Nature's perfection. The *señorita*'s artifice cloaks a veritable statue of Venus, the envy of every sculptor.
>
> Her only feminine weakness is her winsomeness. Headaches, convulsions, and the cerulean pallor of our slow European asphyxiation are unknown to her. Wind and sun caress her skin, and her breasts, ordained by Nature as man's first refuge, have felt no tremor but in her swift, violent gallop across the plains. They are a fruit whose fragrance and form have been cultivated to perfection, but whose savor retains its pure, unadulterated rustic tang. . . .
>
> The Argentine woman generally raises her children herself. She dominates men with her beauty's enchantment, and demands much—because she also has much to offer. With no other argument than her own nature she refutes those European novelists who have attributed to their Creole heroines an excess of sensual appetite. For concupiscence is more the bastard fruit of corruption than the product of strong sentiment, and robust nature is by far more innocent than disorderly impotence.
>
> The Argentine woman also plays an important, albeit indirect role in the affairs of her nation, for her men place great value on her fleeting smile. She will doubtless learn to make better use of this powerful influence than she has to date, softening customs and lending greater weight to genius and knowledge than to riches. (Mantegazza 1876, 45–47; qtd. in Holmberg 1882, 130ff, n. 47)

For Holmberg, Mantegazza's paean to the *porteña* serves to illustrate the transformative potential of female beauty in human evolution, as governed by sexual selection. Like Alberdi, he recognizes the beauty of Argentine women as a lure for foreign men. Unlike Alberdi, however, he does not believe that the evolutionary transformation of the population will leave the original Spanish racial type intact. Like other aspects of the Argentine race, it, too, may be expected to improve.

> As the Province of Buenos Aires has a greater foreign population than the rest of the nation, and the Capital one greater still, the *porteña* exhibits visible change from day to day, such that the time approaches when her grace and beauty will call for another Mantegazza to sing the praises of this marvel. Here there is no longer any single archetype—there is only something indefinable, but precious. In some provinces of Argentina, the ethnic Hispano-Arabic features are retained with some degree of purity, occasionally even recalling one or another particular Spanish type. But they will doubtless disappear, just as they are now vanishing in Buenos Aires, and to a lesser extent, in Rosario. (Holmberg 1882, 131)

Holmberg also recognizes, with Darwin, that sexual selection among humans is driven not by beauty alone, but by a range of confounding cultural factors. An individual devoid of beauty may still enjoy reproductive success, if sufficiently wealthy. In Europe, the prevalence of dowries had largely obviated the evolutionary impact of aesthetic criteria—but not in Argentina.

> Some European nations continue to practice the dowry. A dowered woman has a 99% chance of finding a husband, and since the latter's attention is fixed on the dowry, the selection of physical traits finds itself partially thwarted by a vigorous opponent: money.
> While we here lack the *institution* of the dowry, the *practice* is certainly not prohibited, nor does the possession of a fortune prevent a woman from claiming great personal advantages.
> What's more, thanks to the intervention in our nation of the balancing of wills and reciprocal love as a powerful mediating agent in the selection of beauty, intelligence, and grace, the product of this selection is a beautiful, intelligent, and gracious type.
> We who are ugly are the confounding element. (Holmberg 1882, 131)

As in most treatments of sexual selection by Spanish American intellectuals, the issue of female choice is not addressed. Holmberg's presupposes the passivity of women in the evolutionary process—though he recognizes, with Mantegazza, that a *porteña*'s beauty may facilitate her acquisition of social and political power. Women are here represented as the womb of the future nation, the source from which the future race will spring. Female powers of attraction explain why those more favored by natural selection might choose those who belonged to less favored populations. In this sense, women are empty vessels; unlike men, they do not compromise future evolutionary outcomes.

The two Círculo Médico lectures provide a wonderful snapshot of the reception of Darwinism in Argentina at the time of Darwin's death in 1882. They are by two leading intellectuals of two different generations. One of them was struggling, late in his life, to reconcile the Darwinism to which he had wed his own political fortunes with the pre-Darwinian ideology on which he had made his reputation decades earlier. The other, a boy when *Origin* was published, had been weaned on Darwinism. Born into the Argentine establishment, he had been encouraged, since his youth, to think through the ways in which science might be harnessed in the service of his class and his nation. Both Sarmiento and Holmberg freely and self-consciously exploit the open-endedness of Darwinian analogies, using science to address social and political issues in ways that subtly transformed the meanings of scientific concepts. Natural selection had *naturalized* the extinction of the Indians, granting a plenary indulgence for their slaughter. Sexual selection naturalized beauty, restoring faith in progress by way of ideal types. But, though outside the scope of this book, it is important to mention that by the 1880s this view would also cause conflict regarding the future place reserved for Argentine men. Important debates ensued on the idea of masculinity and the importance of finding such a place. The fear of neurasthenia and degeneration among men, as discussed above, is also related to the interaction of evolutionary views with gender (Novoa 2007; 2008).

In the 1880s Sarmiento and Holmberg were far from alone in ascribing an evolutionary function to female beauty. The discourse of sexual selection was also clearly in evidence in an 1883 piece by Vicente Quesada, once again penned under the Victor Gálvez pseudonym (see chap. 4) reserved for nostalgic reminiscences. In this account of the British invasions of Buenos Aires of 1806 and 1807, Quesada takes up the importance of *porteña* beauty in the conquest and retention of men who, in their colonial adventures, were generally reluctant to mix with exotic strangers.

According to Quesada, the invasions brought "a strange and new element to society, because those distinguished officers, white and blond, with blue eyes, were a novelty for the female population, accustomed to the black hair and the penetrating gaze of the Spanish and *criollo* races."

Local women found these differences intriguing, and so "hearts spoke, and even though they hated the conquerors, they felt sorry for the imprisoned officers, who had been ordered billeted among their families." Upon their release, "the English took with them memories" of the Argentine beauties, so that "the English press focused on Buenos Aires and Montevideo, while a thousand publications, and even the trials of the invading generals, made England trendy throughout the Río de la Plata" (V. Quesada 1883b, 245). When the conflict was settled, English officers unwilling to be separated from the women they had met returned to Buenos Aires, where they contributed to the emerging nation. "From this agglomeration of doctrines, languages, and races, a strong and vigorous race may arise, and at the future's mysterious horizon, it is to this race that destiny belongs" (247).

References to sexual selection as a tool for the deliberate creation of a new racial type, whiter and more modern, may also be found throughout the sociological literature of the 1880s. In an 1888 report on present conditions in the coastal province of Santa Fé, a region that, like Buenos Aires, had absorbed a large immigrant population, sexual selection is explicitly tied to progress. This process is aided by the fact that male immigrants vastly outnumber female; in 1887, "among the 84,215 foreigners [living in Santa Fé], there are only 27,824 women . . . only a third of the foreign population are women" (Carrasco and Ballesteros-Zorraquin 1888, 24). This disparity is viewed as beneficial, because competition for the women magnifies the effects of sexual selection, further improving the emerging race.

> The preponderance of the masculine sex, together with the fact that respect is accorded almost exclusively to foreign men, explain the formation of a new and most powerful race, product of the replenishment of the blood of the American woman with that of the European immigrant, confused with each other by selection.
>
> In the province of Santa Fé, the old type of the Argentine gaucho is rapidly disappearing. There are districts, like that of Las Colonias, in which the poncho and *chiripá*, once our national dress, have become a myth, for they are no longer seen.
>
> Unions between European men and Argentine women are very common, giving rise to new families and contributing to population growth. (22)

Argentine women are the essential breeding stock for the new race. Once again, though, we see them described as essentially passive receptacles for the European infusion. It is European men, and not Argentine women, who contribute the truly desirable traits.

In Argentina as elsewhere, sexual selection added a new dimension to discussions of human sexual dimorphism. These discussions were further complicated by the fact that, in Argentina, as Alberdi's text shows, the presumptive right of local men to mate and reproduce had been downgraded so as to favor foreign immigrants. Yet mate choice itself remained a male prerogative. As for the most desirable male traits, the conditions of natural and sexual selection in Argentina made them a scarce resource, to be managed carefully. The process Haeckel had called "military selection" might well undermine the new race in its infancy, as Holmberg noted in his Círculo Médico lecture:

> As Haeckel observes, defective men, whether on account of their extreme weakness or for other reasons, are of no use in war.[4] The large, the robust, these are of use; these are cannon fodder.... It is they who must expose themselves to bullets, while the others are good for nothing but reproducing themselves!
>
> Military selection, as Haeckel calls it, has not been practiced here to the same degree as in Germany, but its effects may still be observed.... It is nearly certain that in measuring the average height of *porteños* of the third through fifth generations—those whose fathers and grandfathers were also *porteños*— we would find them shorter than the men of other provinces of the Republic.
>
> It's understandable. We are so fond of war!
>
> And as war continues to entertain us the world over, humanity degenerates; those who are in the worst of shape are useless for soldiering, but do a fine job of reproducing, and so the degeneration proceeds, and diseases increase. (Holmberg 1882, 128)

Guillermo Rawson (1821–1890), now the nation's leading hygienist, expresses similar concerns in an 1884 piece.[5] Rawson had been in Europe in the 1870s and had witnessed the aftermath of the Franco-Prussian War. Even during the relatively peaceful 1880s, he observes, military selection continues to operate.

> In 1882 ... with Europe entirely at peace, its soldiers and sailors numbered some 4,140,579.... These 4,000,000 are selected from among the best, by age, health, size and vigor. They are separated from the rest of the population and removed to bases or to the borders to be trained, until it is time to fight. This group of individuals, in the prime years of their life for labor and productivity,

is sequestered together under conditions naturally favorable for the intensification of disease . . . and so the very segment of the population which, under normal conditions of social life, would exhibit the very lowest rate of mortality, perhaps seven in a thousand, must now, under military discipline . . . suffer an average mortality of twelve in a thousand.

What's more, these life forces thus torn from production and the nation's store of assets, are also removed from the family. These strong youths, capable of sustaining their own energy and of transmitting it, in the home, to their children, gifting them with the vitality of their progenitors, are temporarily condemned to celibacy. This onerous deficiency adversely affects the constitution of the family. Paternity is no longer the prerogative of the young and strong. Those rejected as too old, too diseased, or too deformed must take their place, and thus begins, unconsciously, the physical degradation of the race in generations to come. (Rawson 1891a, 214)

Military selection was harmful in part because it counteracted the beneficial effects of natural and sexual selection, of which human sexual dimorphism was one. In another 1884 essay, intended for delivery at the Hygienic Congress in The Hague, which Rawson was unable to attend, the evolution and social function of this sexual dimorphism are explored in detail. Male brawn and intelligence and the female childrearing capacity are the products of *natural* selection; the fact that each is, by and large, the province of one of the two sexes is a function of *sexual* selection (Rawson 1891a, "Diversa resistencia vital de los sexos," 279–91). Natural selection had thus gifted our species with the virtues proper to both sexes, and sexual selection had ensured their segregation, enforcing a socially beneficial division of labor. Military selection removed some of the strongest, brightest men from the breeding population. The socially beneficial aspects of sexual dimorphism would surely suffer.

Argentine criminologists also reflected on the mechanism of sexual selection, blaming it for the perpetuation of weaker individuals whose biological continuity rested on the attraction held by degeneracy among those predisposed. In chapter 3 we mentioned Luis María Drago's *Los hombres de presa* (Predator Men), published in 1887, a great success everywhere the work of Lombroso was known. Drago calls sexual selection "monstrous selection" because, by their "mysterious chosen affinities," the "victims of the same vices and same misfortunes" will be gathered (Drago 1888, 175–76). In his view, this selective mechanism was responsible not for the achievement of higher goals, but for the continuity of those aspects of society that might potentially destroy the nation, such as alcoholism and crime.

The Catholic opposition, vocal as it was, could hardly have remained silent on the subject of sexual selection. As we have noted, the first Assembly of Argentine Catholics in 1884 was largely devoted to denouncing Darwinism. At this gathering, the doctrine of sexual selection came in for its share of attacks. In a fiery oration, Manuel Pizarro insists that civilization consists not in the ability "of Argentine mothers, contemplating the physical strength and beauty of their children's bodies, to exclaim, with the mothers of Greece, *we alone bear men*; but instead, rather than hurl the weak and malformed from the summit of Taigetes, to exclaim, with the Savior of the world, *Oh, suffer the children to come unto me!*" (Asociacion Católica 1885, 61; emphasis in original). The implied reference to the pagan practice of infanticide does not, strictly speaking, target a form of sexual selection, but, like sexual selection, it involves choices made on the basis of aesthetic criteria. Pizarro's disdain for sexual selection extends to the aesthetic project articulated by Sarmiento and others. He excoriates Darwinism as extolling "Greek art's idolatry of form over Christian art's reverence for sentiment and idea." The population policies in which the offending aesthetic principles are implemented "aim to fix the type of our civilization, and are daily becoming maxims of government, as their spirit penetrates legislation and jurisprudence." But his strongest invective is reserved for what Darwin calls the "social instinct," on which Pizarro quotes from chapter 3 of *Descent of Man*, concluding, "Allow me to sum up in a few words this cruel, impious theory, which denies human liberty and sacrifices the bounds of conscience to force. Force, it asserts, is the *law* by which evolution gives rise to *social animals* and thence, by further evolution, to the *intellectuality* of the rational being, who acquires speech and the ability to calculate utility until finally, by further development of animal instincts, he acquires conscience and moral being" (Asociacion Católica 1885, 36). Pizarro has clearly taken the trouble to read and understand Darwin. He is attacking not a straw man but the actual, substantive project of chapter 3 of *Descent*: naturalizing normativity in general and moral norms in particular. With such an approach, morality is seen as a matter of biology or psychology, rather than the spiritual discipline of Catholic Christianity. If improving the race by Darwinian means is to be the overarching goal of public policy, it leaves no place for the Christian mission of caring for the meek. Correspondingly, when grounded in natural and sexual selection, human morality leaves no place for altruism, and thus no place for Christian charity.

Pizarro thus sees the social consequences of Darwinism as both anathema to the Christian faith and a threat to the institution of the Church. Significantly, his critique is couched in just such consequentialist terms. He raises no objections to the hypothesis of common descent per se and never directly considers the question of the phylogenetic relationship between humans and apes. Pizarro's response to Darwinism, like that of José Manuel de Estrada (discussed in chap. 1), is grounded not in the authority of scripture, which he rarely quotes, but in an analysis of the social evils to which the materialism of the dominant national ideology is likely to lead.

Sexual Selection, Aesthetics, and Racial Types

As we have already discussed, sexual selection played an important role in discussions of race throughout the 1880s. Embraced by new standard-bearers, such as sociologist Ernesto Quesada, son of Vicente, its prominence continued into the 1890s. Ernesto Quesada was born in 1858. Like Holmberg, he displayed the strong Germanophile tendencies that had replaced the Francophile proclivities of the prior generation of the Argentine intellectual elite. He completed his studies at the University of Berlin in 1880, after which he held numerous political and diplomatic posts, along with the first Professorship of Sociology in the Faculty of Philosophy and Letters of the University of Buenos Aires. When he died in self-imposed Swiss exile in 1934, his vast personal library became the cornerstone of the Ibero-Amerikanisches Institut in Berlin (Pyenson 2002, 241).

In the late 1890s, Quesada lectured on Darwin to sociology students, expounding an approach to sexual selection quite different from the others we have seen in this chapter and much less optimistic. In a lecture entitled "Herbert Spencer y sus doctrinas sociológicas" (Herbert Spencer and his Sociological Doctrines), he asserts,

> Sexual selection proceeds either by battle between males, or by their display of seductive traits; in either case, the courted female goes to the victor. Thus explicated, the principle cannot be universal, for it cannot be applied to inferior species, demanding as it does a degree of instinct or intelligence, nearly an aesthetic sense, such as only superior species possess. This is certainly true for the second form [of sexual competition], in which seductive traits are displayed, such as brilliant plumage.... But in this respect Wallace diverges [from Darwin], demonstrating that even if such competition occurs, the vanquished males will still, in the long run, find females who take pity on them, allowing

them to reproduce. It follows that the female who takes pity on the inferior male must always be an inferior female, so that the coupling between superior beings always produces offspring superior to theirs. Sexual selection is thus most effective when it operates to slowly accentuate the superior qualities of some, and the inferior qualities of others. (E. Quesada 1907, 14)

Sarmiento and Holmberg had lit upon sexual selection as grounds for believing that the Argentine nation might not, after all, be destined to extinction; properly managed, by means of the cultural impact on sexual selection, it might give rise to a new, better race. Quesada takes Darwin in a very different direction. Sexual selection now serves to perpetuate or accentuate differences between groups. Darwin had opined that members of higher social classes were more attractive, given that their men could always secure the most attractive mates—but Quesada takes his argument much further by considering the fate of those inferior males who, while not entirely without reproductive opportunities, found themselves disadvantaged in mate choice. In doing so, he introduces *pity*, a concept that plays no role in Darwin's account of sexual selection, as a kind of complement to the aesthetic sense. *Female selection*, by which mate choice is often primarily a female prerogative, as in Haeckel's and Darwin's account of sexual selection, is here reduced to female pity.

Sexual selection could not, according to Quesada's view, be taken as underwriting the deployment of beautiful Argentine women as an army for the pursuit of racial improvement. Like Wallace, whom he cites, Quesada viewed "sexual selection [as operating] within natural selection" (E. Quesada 1907, 15). It did not warrant the kind of quasi-Lamarckian return to soft inheritance propounded by Sarmiento. Natural selection demanded *hard* inheritance, but as Quesada notes, the laws and mechanisms governing heritable variation are as unknown at the end of the nineteenth century as they had been to Darwin in 1859. Given this limitation, no demonstration of Darwinian evolution in the biological realm can ever be wholly compelling, depending, as it must, "either on an infinitesimal analysis of the components of organisms . . . or on a very lengthy induction, when the principles are applied to vanished species" (27).

But in Quesada's view, this limitation did not undermine the explanatory power of Darwin's theory. Quesada recognized the theory-constitutive function of open-ended analogies in evolutionary theory and self-consciously exploited it. His understanding of the generality of variation and selective retention allowed him to abstract evolutionary processes from the biological realm. Thus abstracted, these processes could be identified, on the basis of historical evidence, as operating in the sociological realm, where their true explanatory value could become apparent. "It is not biological phenomena, but precisely social phenomena, as Spencer so brilliantly intuited, that have proved most amenable to the demonstration of the Darwinian law . . . variation, adaptation, and inheritance may be examined in social phenomena, so long as history provides us with the necessary elements of the past, as observation furnishes those of the present, without recourse to any hypothesis [concerning unobserved mechanisms]" (27).

Like Sarmiento and Holmberg, Quesada saw Darwin's failure to integrate his theory of evolution into a philosophical system as a limitation, though Darwin might be excused, "because he was a specialist, who never pretended to be concerned with the origins of the universe or life, nor with constructing a philosophical system or cosmogony, but simply with establishing a novel hypothesis" (12). But such a system is needed, lest science succumb "to the intolerantly materialist consequences inferred by those who pretend, following Haeckel, that everything may be explained by eminently mechanistic Darwinian monism" (17). A return to the past, to systematic *Naturphilosophie*, is not an option, for, "in the Germanic countries, Darwinism has provoked a mechanistic, materialist tendency, while in England its impact proved exclusively biological, and not a little agnostic, substituting the genetic principle of development for the morphological principle of mere classification, whose supreme expression was Humboldt's *Kosmos*. This work concluded the old period of science, and Darwin's *Origin of Species* announced the new, removing science from the exclusive atmosphere of the laboratory, to expand its horizons in wide-open nature" (18).

Unlike Holmberg, Quesada finds no help in Haeckel. Instead, like Sarmiento, he turns to Spencer, thanks to whom the "Darwinian doctrine of natural selection has become an integral part of the general theory of evolution, along Spencerian lines" (21). The *generality* of this theory, with its broad applicability to human history, is entirely due to Spencer. Darwin's and Spencer's contributions were complementary.

> Darwin's formulation is confined to a special, limited compartment of science; it owes its philosophical character, its generalizing extension, its essential modality, to the doctrine of Spencer. Spencer had discerned its power and proclaimed it, but in vague terms, more inductively than demonstratively; while

Darwin, unconcerned with its philosophical reach, had expressed its scientific precision by means of technical demonstration. And so Spencer adopted Darwin's formulation, incorporated it into his own system, and enlivened and extended it, converting it into the powerful philosophical instrument of which every branch of human knowledge has now successively availed itself.... So it is via Spencer that Darwin's doctrine has triumphed in the philosophical world, by virtue of its immediate application to sociology. (24)

Quesada is as committed to the future of the Argentine nation as any of his predecessors, but his commitment is to *social* progress, as divorced from any reduction to *biological* progress. Such progress may be understood as, in a sense, evolutionary; but the sense in question owes more to Spencer than Darwin. Where others turned to sexual selection as a way of bringing cultural influence to bear on biology, Quesada was able to treat culture as an autonomous realm, albeit one governed by "exactly the same laws" as biology (22).

If social and biological entities were governed by the same laws, this was because, as Spencer had opined, organisms and societies were, ontologically speaking, the same sorts of things. The analogy between societies and individual organisms is a very old one, indeed, but in Argentina by the end of the nineteenth century, its entailments owed more to Darwin and Spencer than to Plato. Nowhere is this more evident than in a speech by one Representative Vivanco, delivered to the Congress on September 7, 1896, in which he argued against a proposal that foreign schools be obligated to provide instruction in Spanish. Though Vivanco mentions neither Spencer or Darwin, their presence is a given.

> In a simple, irritating, nigh insignificant way, this project attempts to solve a fundamental problem of sociology. It seeks the past, present, and future unity of the organism now in the process of formation, which we call the Argentine nationality, but without taking into account the fact that it is now in the chaotic formative stage, in which different pasts, extant atavistic hereditary influences, and diverging ideals and passions all blend in a boiling fusion. These are the factors that old Europe sends us, together with the native energies of its race, in releasing its excess working population into our immense territory. From these influences, from these germs, a human composite must form, the definitive type we will call the Argentine of the future, in which ... not too long from now, we will supposedly find this complex of sentiments all united, together with beauty and physical vigor, under the banner of patriotic fervor. (N. Carranza 1905, 384)

Another Argentine contribution to the study of sexual selection in the 1890s came in the form of William Henry Hudson's monograph *The Naturalist in La Plata*. Among naturalists, Hudson built his reputation as a field ornithologist, in which capacity he sparred with Darwin in print. This fact is significant, as ornithology had provided Darwin with the most important examples of sexual selection. In this field, at least, if not in sociology or other sciences more narrowly concerned with humans, we find some willingness to entertain the possibility of female selection—though Hudson is skeptical. As of the publication of the third edition of *The Naturalist in La Plata* in 1895, he asserts, the reality of sexual selection remains unproven: "Whether the glittering iridescent tints and singular ornaments for which this family [hummingbirds] is famous result from the cumulative process of conscious or voluntary sexual selection, as Darwin thought, or are merely the outcome of a superabundant vitality, as Dr. A. R. Wallace so strongly maintains, is a question which science has not yet answered satisfactorily" (Hudson 1895, 208).[6] In general, Hudson is confident that future research will lead to "a conviction that conscious sexual selection on the part of the female is not the cause of music and dancing performances in birds" (287). By contrast, as we have noted, those who used analogies derived from the study of mating displays in birds toward understanding humans tended to believe Darwin's account. This emphasis persisted in future years. 1907 saw the publication of the second edition of José Ramos Mejía's *Rosas y su Tiempo* (Rosas and His Time). Ramos Mejía was born in 1852, the year of Rosas's ouster, and would be acknowledged as one of the most influential thinkers of his generation, especially in the field of psychology, to which he devoted his greatest efforts. In this study, Ramos Mejía attempts to explain Rosas's rise to power, and especially his tremendous popular appeal among members of the inferior races, by recourse to a Darwinian mass psychology in which sexual selection played a key role.

According to Ramos Mejía, at the time of Rosas's rise, there was "a kind of battle underway among the [Afro-Argentine] plebes, in which males sought to gain the affections of females; this was sexual selection disguised as political struggle. The symbolic colors in the dress of both sexes, grotesquely arranged and combined, evoked the plumage of birds and the attitudes of other animals, awakening feelings of superiority founded in a polychromatic beauty that accentuated dimorphism" (Ramos Mejía 1907, 429–30). Here we find an Argentine thinker considering the prospect of female selection in humans—but significantly, this possibility is seriously entertained only for the "inferior" races of the "plebes." Indeed, the Afro-Argentines were inferior, in Ramos Mejía's judgment, precisely because their women dominated the men, making them fight amongst each other for their favors.

In Sarmiento's vision of civilized life and the transition to it, it is the women who seduce and the men who choose. Ramos Mejía's portrayal of an inversion of the civilized order rests on an inversion of gender roles in sexual selection. Rosas was able to dominate this inverted order by virtue of his own embodiment of "sexual dimorphism . . . which suggested a cult of protection all the more seductive and interesting because . . . unlike other *caudillos*, who had fathered entire villages, he was, as is well known, a cold man: contained, cruel, and olympically chaste." His followers exhibited a "feeling of sexual orientation proper to butterflies and snakes" (Ramos Mejía 1907, 430) and diametrically opposed to the healthy sexuality of civilized men and women.

Ramos Mejía's text is one example among many in which, in discussions of the politics of mating, the legacy of several decades of Darwinian discourse becomes apparent. Others may be found in the writings of Carlos Octavio Bunge, especially in his works of fiction. Female perversion, which leads, in turn, to male degeneracy, is a common trope in such discussions, running throughout Bunge's dialogue "La perfidia femenina" (Feminine Perfidy). This text presents a conversation on the subject of women between three friends, described as professors of literature, law, and medicine. All three are skeptical concerning the capacity for self-sacrifice traditionally ascribed to women, and though their views are differently nuanced, they all tend to agree with the following pronouncement by Murriondo, the physician: "The biological synthesis of my views on sexual psychology may be summed up in a few words. Three conditions predominate in women, by virtue of their inheritance and physiological organization: a conservative spirit, irritability, and the capacity for fraud. The opposing qualities—an evolutionary spirit, sensitivity, and sincerity—would, then, correspond to men" (C. Bunge 1980, 218–19). The need to maintain sexual dimorphism over the course of a modernization process that was transforming women's roles engendered anxiety about a female threat that might mislead men into following the wrong evolutionary path. When women are understood as cultivating beauty as a weapon in their struggle to reproduce—a weapon wielded most effectively against sincere, altruistic men—female psychology becomes fundamentally deceptive. True, it might be harnessed as a tool in the construction of a new race, but the positive evolutionary contributions to this race must come from men.

The human contribution to the process of national racial improvement was not exclusive to the human species. In a contribution to an official anthology on the Argentine horse published in 1900, Desiderio Bernier discusses the development of a national equine race in terms closely akin to those employed in discussions of the human population. In the view of the Argentine Ministry of Agriculture, the Argentine horse "is the product of the environment into which it has been placed; or more precisely, it is the combined result of heredity and the *struggle for life*" (Bernier 1900, 220; emphasis in original). In this process the "weak has fallen in this fight for existence. The triumph has been on the side of the strongest; and, as a consequence, a *natural selection* has occurred." This process of selection explains the "sobriety, rusticity, endurance, strength of constitution," that characterized this breed. But despite this progress, the Argentine horse remains far from being "an excellent war horse." In order to "correct his defects" it is important to decide to proceed either by "mixing or selection" (221). As in the treatment of human races and their miscegenation during the last part of the nineteenth century, selection is understood as crucial to evolutionary destiny. But in continuity with idealist morphology, civilized activities, such as selection, are related to the realization of aesthetic ideals. Sexual selection, and the creation of ideal types through such selection, bridged the gap for the improvement of humans. Artificial selection was the parallel mechanism for other animals.

Directed evolution, according to this view, was the process of perfecting a type by instilling harmony and beauty. It is for this reason that by choosing "good types as fathers, the race will be modified slowly, until the characters that constitute beauty become fixed in the individual." But attempting to improve a type through selective mechanisms implies a degree of civilization that, as usual, is presented as lacking in Argentina.

> The Argentine breeder is impatient; he wants to observe from one day to the next the fruits of his work. He does not have the resiliency and tenacity that characterize the English breeders....
>
> We admire the creations of others. We are enthusiastic about foreign products, and we do not do anything to make our own. This is the case with the Creole horse, a true diamond that with a little polishing could become an animal of real value, an easy and profitable sell. (223)

Desiderio Bernier reiterates this same complaint in another compilation by the Ministry of Agriculture published two years later, in 1902. The problem with the national attitude is presented in aesthetic terms. Because the current breed of Argentine horse is perfectly adequate to its task, the problem was not its productivity, but its artistic limitations. It lacks "elegance" of form, which meant that it poorly represented the cultural values of the nation. This problem, in turn, resulted from the fact that the improvement of the breed had thus far depended on hybridization alone, an incomplete method. The fruits of this process were "animals without proportion, that seem made by the assembly of parts from the most diverse breeds, reunited and grounded without art, without harmony." The main problem with relying on mixing was that it created a population "without aesthetics." This aesthetic deficit is further related to "anarchy, with all its evils, all its horrors" (Bernier 1902, 104). Coincidentally, the specter of these same horrors looms in discussions of the heterogeneous local human population, responsible for the chaos and social diseases of the beginning of the twentieth century.

The reception of Darwinism in nineteenth-century Argentina took a new turn with the increasing prominence of sexual selection. The great political and cultural significance this notion would acquire in the 1880s and 1890s owed much to the pre-Darwinian articulation of the social significance of female beauty and its deployment, as far back as the 1850s, as a tool for addressing the nation's population problem. It also owed much to the open-endedness of Darwin's analogies. In the hands of some thinkers, this openness sanctioned an attempt to shape the biological fate of the nation, to mold the race of the future, by manipulating aesthetic sensibilities and thus directing reproductive decisions. This attempt gained further impetus from its apparent compatibility with dearly held Enlightenment and Romantic ideas concerning the central importance of beauty in the progress of civilization.

For other thinkers, the open-endedness of the constitutive analogies of Darwinism underwrote an ambitious generalization of the Darwinian account of evolutionary change and its application to a social and cultural realm now conceived as at least partially autonomous from the biological. But of course the constitutive analogies of Darwinism are not *entirely* open-ended. Natural and sexual selection both proceed by the selective retention of chance variations, for whose original emergence and subsequent heritability there was no agreed-upon explanation. Such explanations would eventually be forthcoming—but even when viewed as subject, ultimately, to deterministic laws, variation itself was not *predictable*. The entities governed by Darwinian evolution, whether biological, like species and races, or social, might exhibit one or another sort of *progress* over the course of their evolutionary transformations. But such progress would always be contingent on the particular variations that had arisen over the course of evolutionary history. Evolutionary progress could *not* be understood as a matter of historical necessity; at the end of the nineteenth and beginning of the twentieth centuries, historical teleology was thus deeply problematic.

CHAPTER SIX

Evolutionary Psychology and Its Analogies

Thus far we have explored the impact of Darwinian analogies in politics, paleontology and geology, and, albeit briefly, the social sciences. Darwinism's reach in the social sciences went much deeper than the discussion of the preceding chapters might suggest, in large measure because of the emergence in late nineteenth-century Argentina of the influential schools of evolutionary psychology to which we now turn. These schools did not appear *ex nihilo* with the advent of Darwinism. Instead, Darwinian analogies were integrated into preexisting conceptual structures, the product of an intellectual tradition in psychology that dated back to the early years of the nineteenth century. For this reason, the emergence of *evolutionary* psychology in Argentina serves as a particularly useful exemplar of peripheral science. The mutual accommodations between pre- and post-Darwinian elements transformed both.

The incorporation of psychology into evolutionary discourse coincided with the Anglo-Saxon ascendency in Europe, punctuated by the Prussian defeat of France in 1870–1871. This event profoundly affected Argentine intellectuals, reinforcing the image of the degeneracy of the Latin race to which they, like most Latin American elites, typically belonged. The economic superiority of Britain and the emergence of Germany and the United States as world powers coincided with the rise of Aryanism.[1] Intimations of racial degeneracy gained additional force during the economic and political crisis of 1890, which dashed the burgeoning hope for continuous progress under stable political institutions. In the elite view, the swelling power of the masses, which now took to the streets in defense of new political ideologies, betokened a social degeneracy ultimately traceable to Argentina's racial incompatibility with true civilization.

Because Latin American nations had vested their future in science, it would be difficult to overstate the extent to which this preoccupation with the degeneracy and eventual demise of the race came to dominate scientific thought toward the end of the century. As Nancy Leys Stepan has put it, "the three last decades of the nineteenth century saw growing economic competition among nations and the rise of new demands from previously marginalized groups." At the same time, "the optimism of the mid-Victorian period began to give way to the widespread pessimism about modern life and its ills. Anxiety about the future of progress was reinforced by unease about modernity itself." In this context, "'degeneration' replaced evolution as the major metaphor of the day, with vice, crime, immigration, women's work, and the urban environment variously blamed as its cause" (Stepan 1996, 24). Medicine and psychology—and especially social medicine and social psychology—played a central role in the scientific study of degeneration and its causes. The practitioners of these sciences, *qua* scientists, perforce treated their object of study as part of nature, where "'nature,' or the science that as a human practice produces it, does not escape the value conflicts existing in its social surroundings" (34).

The Argentine Precursors to Evolutionary Psychology

Beginning with Cosme Argerich, an empiricist philosophical psychology had been propounded in Argentina since 1808. Argerich subscribed to the "sensationalism" of fellow physician Pierre Cabanis (1757–1808). Cabanis's mechanistic psychology was, in some respects, a precursor to the work of Comte and Taine. But unlike the Baron d'Holbach, for example, Cabanis subscribed to biological vitalism. Though he believed the study of human cognition, morality, and language had to be constrained by physiology, he was not a strict materialist. He viewed humans as a composite of physical and moral principles, with sensation serving as the link between the two (see, e.g., Baertschi 2005). An entire volume of his *Rapports du physique et du moral de l'homme* (Connections between the Physical and the Moral in Man), published in 1802, was devoted to the physiology of sensation. This work exercised an enormous influence on early nineteenth-century medical thought in Argentina.[2]

Besides Argerich, another Argentine champion of Cabanis and Destutt de Tracy was Juan Fernández de Agüero, a liberal priest and leading partisan of Bernardino Rivadavia's reform movement. He held the professorship of philosophy from 1822 to 1827, when Rivadavia's fall forced him into retirement (Silva 1940, 16). In 1824 he published his *Principios de ideología, elemental, abstractiva, y oratoria* (Principles of Elemental, Abstractive, and Oratory Ideology), in which the influence of his French sources was readily apparent. José Ingenieros would later say of this work that it had set out many of the "standpoints accepted by biological psychology and natural philosophy of the past fifty years . . . [as] notions defined within a coherent, unified system" (Ingenieros 1915, 19).

The influence of the Idéologues on Argentine medicine would remain dominant until 1842, thanks to the efforts of Agüero's successor, Diego Alcorta, the author, in 1827, of Argentina's first dissertation in psychiatry. This study of mania reveals extensive knowledge of the work of Philippe Pinel, the French alienist who, along with Etienne Esquirol, revolutionized the treatment of mental illness (Loudet 1971, 21). In this regard, Alcorta's influences were fairly typical of his generation of Argentine physicians, whose curriculum had been set by Argerich (Ingenieros 1920, 154). Alcorta endorsed Pinel's principle that moral therapy always be accompanied by hygienic therapy (Alcorta 1824). This is particularly significant in light of the fact that his favorite student was Guillermo Rawson (see above, chaps. 1 and 5), who would become the leader of the national hygienist movement and, in turn, an important influence on the figures with whom this chapter is chiefly concerned. Another of Alcorta's students was Vicente F. López (1815–1903; see chap. 1), who offers the following précis of one of Alcorta's introductory courses. "Dr. Alcorta taught from the living corporeal organism, not one of Condillac's statues. As he was a distinguished anatomist and physician, his course featured a comprehensive exposition of the human organism. . . . This first portion of the course constituted his metaphysics or psychology, or rather the basis of those sciences, to be elaborated by further explorations of innate *deductive and inductive ideas*" (V. López n.d., emphasis in original; see also Groussac 1902).

We agree with Ricaurte Soler that the "positivist therapeutic orientation of Ramos Mejía and Ingenieros had its roots in the Argentine ideologist philosophy of the first half of the nineteenth century." The Idéologues are "one of the most stable links between the philosophical thought of the period immediately following Independence and the positivism of the end of the nineteenth and beginning of the twentieth centuries" (Soler 1959, 45; see also Ghiodi 1938). Positivists were in part responsible for the abandonment of speculative philosophical psychology. By the 1850s, psychology had become a clinical discipline, devoted to treating the mentally disturbed. Identifying those who required such treatment remained a political exercise, though the political order had been inverted. After the 1852 victory at Caseros, the returned exiles now in power identified madness with Rosas and his sympathizers. In a letter of June 12, 1852, Juan María Gutiérrez (see chap. 4) informed the president of the Sociedad de Beneficencia (Beneficent Society) that the government would henceforth adhere to the policy of relocating the mentally ill to a place "where they will be spared suffering, and prepared for possible cure" (qtd. in Loudet 1971, 33). Not coincidentally, the place set aside for this purpose was a tract of land that had previously housed the headquarters of the *Mazorca*, along with slaughterhouses and a cemetery, known as "La Convalescencia."

Also in 1852, one of the first official acts of the post-Rosas government was to establish the School of Medicine and appoint Dr. Martín García as the professor of medical nosography and general pathology, in which position he and Manuel Arauz, his successor from 1867 to 1891, disseminated theories of psychopathology. The San Buena Ventura Asylum was founded in October 1863, marking the beginning of a "therapeutic system based on scientific prescriptions of an advanced order, aimed at recreating the special therapeutic practice already extensively tested in similar asylums in Europe, thanks to the happy reforms initiated by Pinel." Dr. Uriarte, the asylum's director, reportedly implemented the "moral treatment of the alienated patient along with medical treatment, and induced the patients to work as an important means of therapy" (Cabred 1895, 528). Within the School of Medicine, a unit devoted to forensic medicine, including forensic psychiatry, was established, first as part of the Department of General Pathology and Medical History and then, from 1875 on, as a department in its own right. A Department of Psychiatry proper was not established until 1886, but the Department of Hygiene was founded in 1873, by recommendation of the Hygiene Council instituted two years earlier.[3] The successive reorganizations of the School of Medicine and other scientific institutions took place while the Darwinian revolution was in full swing.

But the early nineteenth-century tradition of philosophical psychology inspired by the Idéologues was not entirely interrupted. The first head of the Department of Hygiene was Guillermo Rawson, who was a student of Diego Alcorta's. A further thread of continuity was provided by the influential presence of French philosopher Amédée Jacques (1813–1865). Though they later clashed on both philosophy and politics, Jacques had been a disciple of Victor Cousin, who had criticized sensationalism for its excessive materialism, arguing instead for a tripartite division of the mind into reason, sensation, and heart (see contributions to Cousin 1985;

Cousin 1828, 34–35). In addition to the journal *La liberté de penser* (Freedom of Thought) and editions of the work of Samuel Clarke, Fenelon, and Leibniz, Jacques published the highly successful *Manuel de philosophie à l'usage des collèges* (Manual of Philosophy for Use in the Schools) in collaboration with Jules Simon and Emile Saisset (Jacques, Simon, and Saisset 1846).[4] This work became a success, "going through nine printings between its initial publication in 1846 and 1883" (Goldstein 2005, 197). Shortly after its first publication, however, Jacques's thinking underwent a significant shift, when he adopted the positivism of August Comte, whose views were antithetical to Cousin's. In the wake of the revolutions of 1848, Jacques chose to leave Europe, along with many other left-leaning intellectuals. Working at various occupations in Montevideo, Entre Rios, Córdoba, and Tucumán, he finally settled in Buenos Aires in 1863, when Bartolomé Mitre appointed him director of the elite Colegio de Buenos Aires. Jacques's tenure at the Colegio de Buenos Aires was brief—he died in 1865—but significant, as it bridged the gap between the Enlightenment scientism of the generation of 1837 and the positivism derived from Comte. In Soler's words, "Here the convergence of philosophical naturalism with pre-positivist socio-political theory becomes obvious." This convergence occurred in the context of a dominant political liberalism whose rallying cries included anticlericalism and secularism. Thus, "secularism, experimental naturalism, and a sense for the 'objective' in both nature and the social were the factors which informed Jacques's pedagogical spirit, and hence paved the way for Argentine positivism and scientism" (Soler 1959, 50).

Imported along with Comtian positivism were the phrenological theories of Franz Gall, who "had argued that if the brain is responsible for producing the qualities of the mind, it should be possible to work out which part of the brain is responsible for each mental character." By 1840 this hypothesis had given rise to a system in which "a person's character was described from the superficial bumps on his skill, on the mistaken assumption that these revealed the well-developed parts of the brain underneath" (Bowler 2003, 139). Incorporated into Lombroso's system of classification, phrenology would rise in importance later in the century, as Argentine scientists became major contributors to the Italian criminological school.

Well before the advent of evolutionary psychology proper, the Argentine psychological tradition was thus complex and, to a degree, contradictory. On the one hand, it was deeply indebted to Cousin's philosophical psychology, which by dint of its spiritualism remained in tension with the prevailing materialism in natural philosophy. On the other, as the nineteenth century progressed, it came increasingly under the influence of Comte, for whom the qualities of the individual mind were co-determined by biology and social environment.[5] In this respect as in others, the pre-Darwinian Argentine intellectual milieu set the stage for the reception of the Darwinian analogies on which evolutionary psychology would be founded. In it we find strains of biological determinism, balanced by elements of vitalism (derived from Cabanis) and spiritualism (derived from Cousin). Regardless of their metaphysical orientation, Argentines engaged in philosophical psychology and political philosophy sought to bridge the explanatory and therapeutic gap between the individual and the social. Intimations of the Spencerian analogy between the individual human organism and the human species may be found, for example, in Alberdi's writings from the 1840s, well before either Spencerian or Darwinian brands of evolutionary theory. The reception of evolutionary ideas in psychology thus had its roots in earlier debates.

Psychology and Darwinism

Juan Agustín García was born in Buenos Aires in 1862. In 1881 he completed his law degree at the University of Buenos Aires, to which he would return as professor and eventually dean of the Faculty of Arts and Letters. He edited the *Anales de la Universidad de Buenos Aires* and in 1890 published *La ciudad indiana*, destined to become a classic study of the causes, as far back as the colonial period, of the Argentine population problem. In 1899 he published his *Introducción al estudio de las ciencias sociales argentinas* (Introduction to the Study of the Argentine Social Sciences), a monograph on the development and present status of several related disciplines (García 1899). According to García, the rise of Argentine psychology could be understood as the latest of five movements by which Argentine social thought had, over the course of the nineteenth century, successively been dominated: the speculative, the descriptive, the naturalist, the positive, and the psychological. By the time García's study was written, in 1899, the psychological movement, the latest "fashion of the moment," was ascendant (84).

In García's view, this proliferation of psychology was explicable both by its antecedents in the Argentine intellectual tradition and by its effectiveness in bringing the tools of scientific analysis to bear on social concerns. For García, its defining authors included Gustave Le Bon, author of *Psychologie des foules* (Le Bon 1895); Gabriel Tarde, author of *Philosophie penale* and *Etudes de psychologie sociale* (Tarde 1890; 1898); Hippolyte Taine, author of *Les origines de la France contemporaine* (Taine 1871; 1875–1893); and finally Scipio Sighele, author of *La coppia criminale* (Sighele 1892). Taine, Tarde, and Le Bon had extended a medical discourse originally developed to explain certain criminal pathologies, applying it to political mass movements in general. Sighele, a representative of the Italian criminology school, worked in close contact with these French colleagues, who had been struggling ever since the uprisings of 1890 to systematize all available knowledge on the collective mentality of the masses. As Robert Nye describes them, their efforts "spawned a new science of collective psychology in which crowd behavior was described as a collective version of individual mental anomalies" (Nye 1984, 180).

As early as the 1870s, the study of the psychology of the masses had given rise, especially in France, to a new type of historical interpretation, one that echoed the preoccupation with the racial degeneracy of the Latins. Whereas Jules Michelet had seen "the crowd as 'the people,'" the revolts of 1870 had changed the academic perception of the masses. Taine began to refer to them as "the rabble" (Rudé 1964, 8). His *Les origins de la France contemporaine* (The Origins of Contemporary France), which appeared between 1875 and 1893, was written in part as a response to the Franco-Prussian war, the Paris uprisings, and the formation of the Paris Commune. Though Taine himself was interested in explaining developments in French society, many of his hypotheses seemed easily applicable to recent Argentine history. According to one of these, as paraphrased by Susanna Barrows, "the sudden transformation of man into irrational savage is caused by the 'laws of mental contagion.' A solitary man can control his bestial instincts, but once he joins the crowd, 'mutual contagion inflames the passions; crowds end in a state of drunkenness, from which nothing can issue but vertigo and blind rage'" (Barrows 1981, 77).

In France, Taine's influence was enormous. As Barrows puts it, "The mixture of science and politics, of determinism and biology, crystallized into the primary intellectual model for Taine and for many of the subsequent luminaries of French conservative and reactionary thought" (91). More significantly, for our purposes, Taine's biological determinism contained a strong Darwinian component. His book *On Intelligence* includes a careful exposition of the (pre-*Descent*) status of evolutionary theory, in both Darwinian and Spencerian variants. In this work, the purpose of history is articulated as the establishment of continuity between past and present by means of the reconstruction of the evolutionary lineages (see, e.g., Taine 1871, 521–25).

Taine's analytical approach appealed strongly to Argentina's intellectual elite, as it contained two essential elements of the liberal cultural tradition established by those who had fought against Rosas for the adoption of recent European ideas: the medical perspective and the fear of the lower classes. At a time when the Argentine elites saw their country as in decline—facing the evolutionary process of degeneracy and eventual extinction described in the previous chapters—collective psychology struck them as an effective weapon in formulating and mounting their response to social disorder. In García's words, collective psychology allowed for the discovery of "the general features of national character, and the laws governing its operation." It sanctioned the study of the "energy of a nation and its influence on government, its origins in the physical conditions of the territory, and its historical and racial precedents, as well as national *sentiments*, with emphasis on the most sociable among them, along with both their natural tendencies and the proper means of educating them, should they as yet have failed to reach their full potential" (García 1899, 84). The proper means of educating the national sentiments were precisely what the governing elites had been looking for all century.

The ascendancy of psychology coincided with the period when, in Europe, the "eclipse of Darwinism" was clearly in evidence. "From the high point of the 1870s and 1880s, when 'Darwinism' had become virtually synonymous with evolution itself, the selection theory had slipped in popularity to such extent that by 1900 its opponents were convinced it would never recover." The idea of evolution itself wasn't abandoned, but a growing number of biologists "preferred mechanisms other than selection to explain *how* it occurred" (Bowler 2003, 233). The crisis of evolutionary thought was exacerbated by debates between neo-Darwinians and neo-Lamarckians, which exposed the weaknesses of current theory. As we have discussed, the work of August Weismann, the German embryologist who had impressed the scientific world by arguing for the relative importance of hereditary factors over social influence, tended to favor the primacy of natural selection over other evolutionary mechanisms (Weismann 1882; 1891–1892; 1893). He had proposed in "the mid-1880s that scientists should assume the existence of an unbridgeable gap between the 'germ' cells and the 'soma' (body) cells, the former being the unique and independent carriers of hereditary information." His ideas were later incorporated "into Mendelian and neo-Darwinian inheritance theory, which triumph virtually everywhere [in Europe] after 1900 except in France" (Nye 1984, 84). In France, by contrast, the epigeneticist view, which emphasized the importance of the social medium in the transmission of hereditary traits, continued to dominate. "French embryologists continued to favor this *epigenetic* tradition well past the time it was abandoned elsewhere in favor of the neo-Mendelian genetics, which held that the hereditary information contained in the chromosomes of the parents could not be influenced in any way by the intrauterine environment" (84).

In Latin America, too, many scientists continued to cling to the neo-Lamarckian views Weismann had destroyed by proving that changes introduced by society and education weren't heritable. As Nancy Leys Stepan has explained it, the Latin American opposition to Weismann was a consequence of the fact that the German biologist's theories "implied . . . a determinism that seemed to leave no place for individual will and action in the development of human society." As we discussed in chapter 3, like their French colleagues, Latin American scientists "could not accept the absolute separation between soma and germ plasm demanded by Weismannians and strict Mendelians." In its political use, neo-Lamarckism "also often came tinged with an optimistic expectation that reforms of the social milieu would result in permanent improvement, an idea in keeping with the environmentalist-sanitary tradition that had become fashionable in the area" (Stepan 1996, 73).

In the midst of this evolutionary debate, José María Ramos Mejía published his *La locura en la historia* (Madness in History) in 1895. Ramos Mejía had studied medicine under Guillermo Rawson, gaining early recognition, even before completing his doctorate, for his *Las neurosis de los hombres célebres* (Neurosis in Famous Men), an 1878 study that anticipated many aspects of later social psychology. He defended his dissertation, *Apuntes clínicos sobre el traumatismo cerebral* (Clinical Notes on Brain Trauma), the following year. *La locura* sported a lengthy introduction by the French émigré scholar Paul Groussac, director of the National Library. It is to Groussac's introduction, which contains a snapshot of many of the then current debates surrounding evolutionary theory, to which we turn first.

Following French intellectual tradition, Groussac strongly criticizes several elements of Darwinism, including some to which Ramos Mejía publically subscribed. Concerned to spare Ramos Mejía any sense of awkwardness, Groussac reports having offered to retract his introduction rather than disagree with the author of the book being introduced. In response, "Ramos Mejía did not merely grant permission; he actively requested, in the name of science, that I make public, without pretense or reticence, my objections to the theories and methods of the scientific school to which he belongs" (Groussac 1895, viii). According to Ramos Mejía's account, madness is a heritable consequence of degeneracy, explained in broadly Darwinian terms. Groussac focuses on "the theory of degeneration," which he calls "a parasitic excrescence of general inheritance that has become dogma with the popular triumph of Darwinism" (xxiii).[6] As you will recall from chapter 4, the preoccupation with degeneracy emerged as part of the national obsession with extinction, which it was thought to presage, where the complementarity of evolution and extinction was unique to Darwinism. Ironically, in Groussac's view, now that Argentina had embraced this brand of evolutionism as official doctrine, the scientific community was abandoning it.

> It would be an arduous task, and a strange one in our view, to determine whether, in the end, the transformist hypothesis had advanced or detained the march of progress. Where once the intellectual plebes rejected it out of ignorance, now they have thoroughly dogmatized it. It takes as much moral valor to dispute Darwinism now as it did, thirty years ago, to defend it publically. The freedom of scientific belief has surely lost its virtue when it becomes entrenched and petrified in intangible dogma.... At the very moment of vulgarization, with the whole flock blindly accepting it, transformism is now losing ground within science. Darwin's conscience, good faith, and sagacity—in short, his genius— are not in question. But no one who thinks for himself can any longer accept all of his bold conclusions "en masse," let alone the gratuitous hypotheses and phylogenetic trees of the Haeckelian sect, whose membership the transformist Claparède has called the *"enfants terribles* of Darwinism." (Groussac 1895, xxiii–xxiv)

Like other critics before him, Groussac charges Darwin with placing too much stock in his analogies. In consequence, "Little by little, Darwin became caught up in the current of his own theory, carried well beyond his primary inductive evidence." Many of the "references cited in support of his doctrine derived from strange sources," some of them hangers-on who "vainly sent their own 'contributions' to the English savant who, being such a prudent and careful observer himself, committed the error of candidly attributing his own virtues to these unknown correspondents, accepting their data without the necessary critical scrutiny." The charge of sycophancy has unmistakable local targets. According to Groussac's account, Darwin had asserted the existence "of a true *race* of ñata cattle; and this extraordinary fact was religiously accepted by Argentine naturalists: *Magister Dixit*" (xxiv; emphasis in original).

In support of his claim that it is no longer possible to accept the entire body of Darwin's conclusions, Groussac cites "the favorite student, personal friend, and companion of Darwin's for fourteen years" (xxiv), George John Romanes. In his 1886 *Physiological Selection*, Romanes had asserted that "at the present time it would be impossible to find any working naturalist who supposes that survival of the fittest is competent to explain all the phenomena of species formation" (Romanes 1886; cited in Groussac 1895, xxiv). More specifically, as James R. Moore notes, "Romanes acknowledged the inability of natural selection by itself to account for phenomena related to the origin and persistence of useless characters; the occurrence of free intercrossing and blending inheritance; and "the origin of cross-infertility between allied species occupying common or closely contiguous areas" (J. Moore 1979, 178).

Indeed, in the late 1880s and early 1890s (he died in 1894), Romanes became an important thinker for those who sought to challenge August Weismann's defense of natural selection as the sole important evolutionary mechanism. Romanes was used by both Darwinians and anti-Darwininians to question the dominance of hard inheritance. For a Frenchman like Groussac, for example, Romanes is of interest as a critic, not a defender, of Darwin, because he retains certain Lamarckian ideas. He cites him as demonstrating "that natural selection is at best a secondary factor in the formation of species and even varieties, much less important than the action of the environment, which the Master so disdained," as showing "what's more, the uselessness of the great majority of species characters, and reaching a conclusion deeply perplexing to the transformist thesis—and its derivatives in the field of psychopathology—that the underlying law is not one of heritable variation, but of reversion to type" (Groussac 1895, xxv).

Groussac proceeds to make similar use of recent work by such apparently staunch Darwinians as Carl Vogt and Thomas H. Huxley before concluding, "the entire theory is collapsing" and "has been razed to its foundations" (xxvii). So widespread is the abandonment of Darwin that, "far from maintaining supremacy, Darwinism appears soon destined to join the large collection of different, and sometimes diametrically opposed, conceptions with which we have tried, in vain, to explain the origins and succession of organic species" (xxvii). Before returning to hereditary psychopathology, the actual subject of Ramos Mejía's work, Groussac offers the following parting shot: "Except for the fruitful notion of evolution, which Darwin popularized rather than discovered, nothing will remain of his general conclusions within positive science." (xxviii). The problem with Darwinism and "many 'scientific' theories, resides in the fact that their method is not properly scientific" (xxix). As Groussac correctly observes, for Ramos Mejía and others, "the theory of hereditary degeneration is the very axis of all the medical literature belonging to the genre of *La locura en la historia*" (Groussac 1895, l), where this is the very theory that Groussac has been at such pains to demolish in his introduction.

La locura en la historia takes hard inheritance for granted, simply assuming, over Groussac's objections, its compatibility with evolution by natural selection. The task of the book, then, is to reconstruct the process of selection in humans on the basis of historical evidence. Of greatest interest for our purposes, and of greatest importance to evolutionary psychology, is not Ramos Mejía's interpretation of this evidence, but his recognition that the task he has set himself demands that he confront, and possibly abandon, the distinction between natural, artificial, and sexual selection. "This providential force operates on the human species, too, but in a slow, laborious way, never achieving the singular promptitude attained by our species itself in its artificial selection of inferior animals. In man, selection is natural, and only the great influences with which we will be concerned [in this chapter, such as the Holy Inquisition], only these agents of curious character, have demonstrably achieved it artificially, by unconsciously, without any explicit goals, eliminating an enormous mass of living, sentient human flesh" (Ramos Mejía 1895, 472–73). In addition to the Inquisition, Ramos Mejía cites slavery as one institution capable of exercising large-scale domestic selection on humans. But what we wish to underscore in this passage is Ramos Mejía's assertion that artificial selection can operate on humans (and presumably on other animals) by way of *unconscious* motives and preferences. These unconscious motives and preferences are presumably also available, in principle, as *targets* of selection, both natural and artificial. Such assertions regarding the unconscious would prove crucial to the next generation of Argentine evolutionary psychologists.

Tracing the evolutionary effects of significant historical events is possible, Ramos Mejía argues, because while subject to natural, artificial, sexual, and unconscious selection, humans in general are evolutionarily quite stable, on account of their capacity to resist all but the fiercest selection pressures. Consequently, "human selection has never occurred, and never occurs now, except at a much lower amplitude than in other animal species, because according to [Adolphe] Quetelet, man exercises a perturbing effect on himself and on everything around him. It is this effect that protects all those who are physically or intellectually too weak or useless from disappearing in the struggle for existence" (475). Like his reference to Weismann on inheritance, Ramos Mejía's allusion to Belgian astronomer and statistician Adolphe Quetelet's pioneering work is nearly prophetic. As André Ariew has shown, despite the later biological significance of Quetelet's work, Darwin had dismissed it (Ariew 2007). In the twentieth century, the incorporation of Quetelet's methods into population biology would radically transform evolutionary theory, marking perhaps the most important difference between Darwinism proper on the one hand, and later neo-Darwinism and the Modern Synthesis on the other. Ironically, it was this transformation that saved Darwin's legacy, allowing population geneticists to finally address such problems as the evolutionary stability of useless characteristics from within the general Darwinian framework of evolution by variation and selective retention. It would clearly be absurd to say that Ramos Mejía saw these developments coming. What can be said, however, is that in Groussac and Ramos Mejía we find a sophisticated, nuanced discussion of the shortcomings of Darwinism and their potential solutions. Defending Darwin, Ramos Mejía asserts that he had "never uttered the phrase, *perfection of a being*, without adding another: *relative to the organic and inorganic conditions of his existence*, or something similar" (Ramos Mejía 1895, 505; emphasis in original).

In any case, Ramos Mejía was largely responsible for Argentina being one of the first countries to develop active research programs in evolutionary psychology. José Ingenieros was his student. Before we take up Bunge and Ingenieros, however, we must turn to the French and German influences that made themselves felt after the decline of Spencer, the ascendance of neo-Darwinism, and the philosophical crisis that these events triggered.

The Emergence of Evolutionary Psychology

Roots and Development

As we have discussed, before the Darwinian revolution, Argentine psychology had begun to crystallize around two very different seeds. The first was the philosophical influence of the Idéologues; the second was the pragmatic, therapeutic orientation of Rosist and post-Rosist medicine. It was to this mix that Darwin's and Spencer's understanding of psychology contributed, fostering new ways of representing the relationship between the biological and the social. For some practitioners, the mental states of individual members of a population became diagnostic of the evolutionary health of the population as a whole, and conversely, when a population was threatened with extinction, the expression of characteristic psychological symptoms was likely to follow. Psychiatry treated the problems of the individual, but in doing so, it addressed the problems of the nation.

Philippe Pinel's work was important in Argentina since at least the early 1860s. In 1876 Dr. Lucio Menéndez instituted "alienism," based on the principles of Pinel and Esquirol, as the official doctrine of the nation's asylums. In 1879 Menéndez, another student of Rawson's, coauthored a statistical study of mental health in the province of Buenos Aires (Menéndez 1879).[7] By 1880 "Argentina's own psychiatric apparatus, and soon her criminological apparatus, [were] under construction, thanks to the creation of new hospitals and academic departments," and also thanks "to the first texts, owed principally to the labors of José M. Ramos Mejía and Lucio Menéndez" (Vezzetti 1985, 49; see also Vezzetti 1988). By this time a number of private institutions had also been founded, including the Instituto Frenopático. Ramos Mejía's book ensured that an account of "the issues of psychopathology achieved such wide distribution as to reach even the community of nonspecialists" (Vezzetti 1985, 48). This community included the political establishment, and so it is hardly surprising to see the theories of mob behavior developed by criminologists reflected in official policies and legislation.

In an 1894 report to the municipal government of Buenos Aires, Domingo Cabred (1860–1929), director of the Psychiatric Clinic at Las Mercedes hospice, discusses the situation of the mentally ill. This Corrientes-born physician had earned his doctorate at the University of Buenos Aires in 1881 with a thesis on "reflex madness." He succeeded Lucio Menéndez as head of Mental Pathology and in 1893 was appointed director of the Psychiatric Clinic (Loudet 1971, 61–65). His report acknowledges the newfound importance of psychiatric research, to which resources were being committed; "The School of Medicine has offered to acquire the instruments necessary to construct a laboratory for the Psychiatric Clinic, and consequently, we may soon count on having all the prerequisites for scientific investigation" (Cabred 1895, 540).

Cabred made another highly significant contribution to the study of psychiatry and related sciences in Argentina. In 1898, as Triarhou and del Cerro recount, he embarked on a European trip with a view toward exploring ways to promote neuroanatomy in psychiatric research back home, a venture he brought to successful conclusion with the hiring of Christfried Jakob (1866–1956) to direct the Laboratory of the Psychiatric and Neurological Clinic at Hospicio de las Mercedes for a three-year term, an offer earlier declined by Adolf von Strümpell (Triarhou and del Cerro 2006, 177). This acquisition was a real coup. Jakob, who "is considered the father of neurobiology, neurology, and forensic histopathology in Argentina," remained in the country after the term of his initial contract expired. A key motive for his relocation "was the prospect of obtaining 300 brains annually for pathological study, when in Germany the corresponding number was 2–3 brains" (177). Where Moreno had collected Indian skulls, Jakob attended to their contents (see, e.g., Jakob 1906).[8]

Jakob arrived in Buenos Aires in July 1899, and for a "dozen years that followed, he produced works in neurology, psychopathology, biology, anthropology and paleontology." The names of many of his collaborators will be familiar from previous chapters, including psychiatrist José Ingenieros, paleontologist Florentino Ameghino, naturalist Clemente Onelli, and anthropologist Robert Lehmann-Nitsche. In his home field of neurobiology, the laboratory he directed enjoyed a reputation as "one of the most important neurobiological laboratories in the world" (Triarhou and del Cerro 2006, 186). Jakob's *Atlas of the Nervous System*, which was translated into several languages, including English, became a standard text (Jakob 1899; 1901).[9] It would be difficult to overstate his impact on Argentine neurobiology and neuropsychology.

Jakob's recruitment coincided, not coincidentally, with the late nineteenth-century Germanophile period noted earlier, during which the influence of Wundt's psychology reached its peak (Wundt 1928; 1897–1907; 1912; 1907; 1897; 1969). Subsequent evolutionary psychology was thus experimentally and, to an extent, neurologically grounded. But because of the mounting influence of Wilhelm Wundt (1832–1920) and his followers, it was also steeped in aspects of the German idealist tradition. We will consider the tensions between this tradition and the biological determinism implicit in neurophysiological reductionism later in this section. But despite this conflict, both biological determinism and Wundtian psychology posed direct and ultimately fatal challenges to the Spencerian metaphysics Sarmiento had so cherished. Evolutionary thought itself, however, would remain a powerful influence. As Mauricio Pappini has noted, during this period the situation in Argentina "differed from that of other Spanish-speaking countries in the relatively important influence of evolutionary biology on the thinking of several of the most influential psychologists" (Pappini 1988, 132; see also Klappenbach 2006, 75–86).

In 1902 psychologist Horacio Piñero explained that three events accounted for the present state of his field. The first two took place in 1878, when "Charcot published his studies of hysteria and hypnosis; and in Leipzig, Wundt founded the first laboratory of experimental psychology." Around the same time, "Ribot founded the *Revue Philosophique*. We can say that this trio of events gave rise to the clinical observation, experimental research, and scientific communication on which rests the valued autonomy of psychology" (Piñero 1900, 117; emphasis in original). Piñero underscores the *autonomy* of contemporary psychology as a particularly significant achievement. Given his emphasis on its experimental component, it is clear that, in clear departure from the Idéologues and Cousin, the issue is independence from *philosophy*. Much of Piñero's article consists in a review of the world's principal laboratories of experimental psychology, and it is a point of special pride for him that, of all the nations of South America, Argentina is "currently the only one that can boast laboratories of experimental psychology, which we have had the opportunity to establish at the Colegio Nacional and at our Facultad de Filosofía, as indispensable elements of the scientific study of this discipline" (133). As a psychology instructor at the Colegio Nacional beginning in 1898 and at the Facultad de Filosofía y Letras beginning in 1901, Piñero had had a hand in both institutions.

Having been dispatched by the government to explore the state of play in European psychological research, Piñero is in a position to provide a firsthand report of recent developments. A member of the International Institute of Psychology founded by Pierre Janet (1859–1947) in 1901, he had networked extensively, attending many international conferences. He takes particular note of an event chaired by Ribot in 1900, at which he observes a marked trend toward neuroanatomy and neurophysiology. This particular conference featured "very, very few" papers devoted to "logical operations, judgment, reasoning, and finally, child psychology, which has taken on a more systematic form, becoming the embryological and genetic study of the human spirit" (134). Piñero's article concludes with a review of recent advances in laboratory instrumentation, new library collections, and new curricula in basic psychology, finally endorsing Janet's call for psychiatric clinics in which the approach to treatment diverges from that in hospitals. Their goal would be to "always address the mental state first, and the physical state second, in search of psycho-physiological parallelism" (138). Janet influenced other Argentines, too, including Jorge Mirey, sometime lawyer, professor at the Escuela Normal de Profesores of Buenos Aires, and author of the 1904 text *Curso de psicología científica* (Course in Scientific Psychology) (Mirey 2004).

Psychology's turn away from philosophy was not universally applauded. In a 1904 report to Norberto Piñero, dean of the Faculty of Philosophy and Letters, Horacio Piñero acknowledges this fact in a review of the current psychology curriculum, addressing the tension between purely experimental psychology and metaphysically grounded philosophical psychology. On being assigned to teach the introductory course, he became aware of the need for *two* courses, one of which offers beginning students the rudiments of experimental and physiological psychology and the other to cover "philosophical, metaphysical, or higher psychology" (Piñero 1904, 391). In deference to the unity of science, he argues, psychology must be taught as a synthetic discipline. While grounded in physical law, it should not disparage either metaphysics or the need for a phenomenological approach to processes inaccessible to experimentation. As Piñero is well aware, psychology, like other sciences in the post-Darwinian context, remains philosophically deficient; recall that it was this deficiency that had made Spencer so appealing. Darwinian analogies are highly fruitful, but in the absence of a totalizing system into which evolutionary theory might be incorporated, the *reason* for their fruitfulness remains elusive.

> To date, psychology remains limited as an experimental science. But such limitations aside, experimentation is not the only way to acquire knowledge, nor are observation and experience the sole sources of insight. Logical reasoning and even metaphysics must furnish us with the explanations for many phenomena that cannot yet be reduced to experimental determinations. This speculative portion of the science, which varies from age to age as new discoveries are made and new theories and conceptions devised, is another worthy object of study. Through reason, metaphysics gains knowledge *of* the experimental disciplines. Metaphysics, the third degree of every science worth calling scientific, is necessary to psychology, as James says, to ensure its future. (392–93)

The second course requested in Piñero's report to his dean was established in 1906. The minutes of a session of the Executive Council of the Faculty of Philosophy and Letters record the assignment "of the first course to Dr. Francisco de Veyga, Interim Professor of Psychology, and the second course to Dr. Felix Kruger [sic], Interim Professor of Psychology" (Facultad de Filosofía y Letras 1906, 323). As Ernesto Quesada described him, Felix Krueger (or Krüger, 1874–1948), had been an assistant in Wundt's laboratory and was now scheduled to arrive in June 1907 along with equipment destined for the Instituto Nacional del Profesorado (National Institute of Secondary Education). The latter institution, founded for the purpose of training future secondary school teachers, was largely organized in accordance with the German model, and intellectual differences with the Faculty of Philosophy and Letters kept the relationship between the two units tense (Klappenbach and Pavesi 1994). Its rise testified to the government's increasing interest in fostering German thought following the French defeat in 1870. This policy brought the government into conflict with Argentine intellectuals, who viewed it as an unwarranted interference in local culture. It did not pass unremarked in Germany, either. Krueger's appointment was duly noted in the 1906 edition of the *Handbuch des Deutschtums im Auslande* (Handbook of Foreign Germany): "The Argentine government has quite recently offered new evidence of the high esteem in which it holds German science, by selecting a student of Wundt's, Leipzig adjunct professor (*Privatdozent*) Felix Krüger to establish a new facility for experimental psychology, granting him a two-year appointment" (Allgemeiner deutscher Schulverein 1906, 352). Unlike Jakob, Krueger did not remain beyond the term of his initial two-year appointment. As Mariano Plotkin notes, his stay was significant, however, in that his teaching "emphasized the nonexperimental aspects of Wundt's work and introduced the works of such nonpositivist writers as Wilhelm Dilthey" and thus proved influential "in triggering the antipositivist reaction of the following decade" (Plotkin 2007, 18).

Following Krueger's stint in Argentina, he occupied various posts until 1917, when he inherited Wundt's chair. From 1927–1933, he was president of the increasingly nationalistic *Deutsche Philosophische Gesellschaft* (German Philosophical Society). As Hans Sluga notes, "As a conservative Krueger had been sufficiently sympathetic to the Nazis to endorse them even before the collapse of the Weimar Republic. In 1933 he welcomed Hitler's rise to power in enthusiastic words. . . . In 1935 he was, in addition, made rector of the University of Leipzig" (Sluga 1993, 226). In 1938, however, he was forced to retire, having publicly expressed some reservations about anti-Semitism. The issue of Krueger's intellectual and political involvement with Nazism is obviously well outside the scope of this study; it deserves mention here only because of Krueger's Argentine impact and because of the interesting parallels between German and Argentine intellectual history. Both cultures struggled to respond to scientific materialism in general and Darwinism in particular. Haeckel's nonreductive monism represented one German attempt to incorporate aspects of Humboldtian *Naturphilosophie* in a Darwinian framework. In its systematicity and teleology, it appealed greatly to one generation of Argentine intellectuals. In Krueger we find a later rapprochement with German idealism. Describing Krueger's mature thought, Sluga notes that, for him, the "concept that linked the fields of philosophy and psychology" was that of "*Ganzheit* [wholeness]." He "was convinced that the mind was by no means an additive unity but that it formed an organic whole" (96; emphasis in original).

Cecilia Taiana has discovered and analyzed the letters Krueger wrote to Wundt while in Buenos Aires. They clearly indicate that much of Sluga's assessment applied equally well to the younger Krueger. But the Argentine appeal of his approach would become evident only in historical hindsight.[10] Despite Piñero's lip-service to the need for metaphysics in general, and philosophical psychology in particular, Krueger's tenure in the Faculty of Philosophy and Letters's second chair in psychology appears to have been a frustrating one; *Ganzheit* simply did not mesh with the largely mechanistic approach that still dominated the field in Argentina, an approach congenial with the work of Krueger's countryman Jakob, but more generally associated with such French figures as Ribot. In a letter to Wundt of November 16, 1906, Krueger laments, "Everybody interested in psychology and related areas is well aware of your name. Unfortunately direct knowledge of your work is very rare. Only the *Grundriss* [*Grundriss der Psychologie*, or *Outlines of Psychology*, an introductory text] has been translated into Spanish, and quite poorly. Prof. Piñero and his students have based their entire scholarship on Paris" (cited in Taiana 2005, 385; see Wundt 1896). So while the Argentine government may have turned its gaze toward Berlin (or at least Leipzig), the Argentine academy remained firmly fixated on Paris, with its positivist proclivities. As Cecilia Taiana explains, one fact that explains the complex reception of Wundt's ideas is the lack of familiarity with the German strands of theoretical psychology among Argentine psychologists. Most of them did not know Wundt's work directly, but only the response he had provoked in Europe, particularly in France (Taiana 2005, 396). In a letter to Wundt of June 11, 1907, Krueger remarks, "Within the faculty [Horacio Piñero] tends to sharply differentiate our respective teaching activities, in the sense that he maintains control over what he describes as 'exact' or strictly scientific psychology, that is, anatomy, physiology, and pathology of the nervous system, and whatever may be actually or supposedly based upon it, while I would be free to lecture on 'spiritualist' or 'metaphysical' psychology. My assignment includes, as you can see, the greater part of empirical psychology, and so it will be quite possible to work next to this ambitious, intelligent, though somewhat pushy man" (Taiana 2005, 386).

In addition to the French influence, Krueger astutely discerned the importance of evolutionary theory in general, and of Herbert Spencer in particular, whose "bold analogies seem to be particularly attractive to new nations (Balkan!)" (Taiana 2005, 386). But despite the philosophical significance of these analogies, "[Horacio] Piñero and the majority of his students consider [philosophy] completely superfluous, even harmful—and this is evident from the curriculum.... Students ... equate psychology with physiology of the brain, and science with natural sciences, [they have] a most confused concept of 'sociology' that is, of course, of French origin. The concept of *Geisteswissenschaften* [Human Sciences] is so little known here that there is not even a generally recognizable term for it. Sciences and exact sciences are exclusively natural sciences" (Taiana 2005, 386–87). As presented by Taiana, Krueger's letters to Wundt offer a fascinating glimpse of the intellectual milieu of Buenos Aires in the first decade of the twentieth century. Viewed in context, they describe a climate that had already begun to change. The urgency with which the latest in European thought was sought out, absorbed, and accommodated to the specific intellectual demands of Argentina made such change inevitable, as Piñero was surely aware when he called for the establishment of Krueger's chair. The Faculty of Philosophy and Letters may still have been largely dominated by French influences when Krueger arrived, but the fact that a collaborator of Wundt's was hired to fill this chair is evidence that interest in his work was already on the rise, just as positivism and Spencerianism were already on the decline.

In 1905 E. J. Weigel Muñoz, an attorney and professor of law at the University of Buenos Aires, published a pamphlet entitled *La psicología y las ciencias sociales*, drawing on Wundt to attack the critical analogy on which Spencer's approach to the social sciences was based.

> The publication of the fundamental principles of Wundt's physiological psychology has shed new light on research in the field known as sociology. As the sage Leipzig Professor has demonstrated, it is a mistake to identify, by seeming analogy, the biological functions of nature with the collective manifestations of groups [of people] subjected to a political regime. The difference between the individual elements of a society (persons) and the histological elements of

a material organism (cells) is not only morphological; the psychic activity of a superior individual also differs in intensity and *modus operandi* from the intentionality of the simulating or dissimulating processes of a cell. (qtd. in Bibliografía 1905, 517)

With Wundt's help, Weigel Muñoz draws attention to the very element of Spencerian sociology that had, for nearly two decades, allowed many Argentine thinkers to incorporate the theory of evolution into a larger philosophical system: the assumption that the conclusions drawn by biologists concerning the evolution and development of species and organisms could be applied to the evolution and development of societies.

This assumption was now under attack by Wundt and others, and the confidence with which Argentine thinkers invoked it was crumbling. Weigel Muñoz was giving voice to doubts that had, by then, become pervasive—though apparently Krueger remained unaware of them. As we pointed out in chapter 3, in a 1907 lecture on political philosophy, we find Antonio Dellepiane expressing similar concerns regarding the "negative and disturbing" effects of the "organicist hypothesis." By dint of their evolved psychological capacities, humans are subject to evolutionary processes that differ in kind from those operating on the lower animals. In humans, for example, "sexual selection is conditioned by far more diverse factors—emotional, economic, political, religious, etc.—than the phenomenon by the same name in species inferior to our own" (Dellepiane 1907, 66–67).

In an argument reminiscent of Ramos Mejía's *La locura en la historia*, Dellepiane charges that advocates of the "organicist hypothesis"—Spencer's claim that the laws of biological evolution govern social evolution—have missed or forgotten something very deep about the constitutive analogies of evolutionary theory itself. In Darwin's original development and exposition of the concept of *selection*, the analogy to the act of *choice*, as undertaken by a human agent, is absolutely central; natural selection, in which human agency is absent, is explained by analogy to selection under domestication, which *requires* human agency. The order of epistemic dependencies is important and must not be overlooked when considering processes partially governed by human agency. Criticizing Italian sociologist Michele Angelo Vaccaro, Dellepiane offers the following analysis.

> Vaccaro has attempted to show that the ultimate social law, to which all others are subordinate, is that of adaptation. Toward this end he parades an enor-

mous array of data before the reader, evidence that the evolutionary process of societies is a result of selection. But as he himself takes pains to show, human selection is not natural but artificial, in that the process is mediated by factors of a psychic order.... Even in his adaptation to the physical environment, man exhibits certain characteristic modalities, in virtue of which his adaptation is substantively different from adaptation in animals. Whereas the latter really do adapt to environmental conditions, undergoing organic change, the former adapts the environmental conditions themselves, creating artificial instruments for his own protection, and employing for this purpose his intelligence, which itself evolves, and is subject to greater transformation than his body. In our opinion, we are thus forced to recognize that the process of human adaptation is one of psychic order. (Dellepiane 1907, 66 n. 1)

Dellepiane's argument deploys the conceptual resources *of* evolutionary biology to make the case for the autonomy of the social sciences *from* evolutionary biology. In spirit, it is strikingly similar to recent arguments for much the same conclusion, including that of Richard Dawkins (2006, chap. 11). Social science, as Dellepiane envisions it, remains the study of evolutionary processes; but the processes in question are not biological. In place of the Spencerian analogy between species, organism, and society, he proposes revisiting the ancient analogy between individual and collective psyche, but "no longer using this and other analogies in the merely metaphorical sense formerly ascribed to them, but to express a substantive reality, the existence of a collective soul or ego, similar to the individual ego and governed by the same laws" (Dellepiane 1907, 76). As conceived by many of the figures canvassed in the preceding chapters, the construction of the modern nation was to be an essentially biological project. Should the biological analogies on which it rested prove inadequate to the task, social psychology might succeed where sociology had failed, with constitutive analogies and metaphysical commitments of its own.

The Spencerian application of the lessons of biology to the social realm had turned in part on Lamarckian presuppositions. Wundt had directly challenged the analogy between organism and society—what Dellepiane called the "organicist hypothesis." At the same time, August Weismann's discovery of the segregation of germ and soma conclusively refuted Lamarkianism. Spencer's approach to the social sciences, as it had flourished in Argentina, had been undermined. Though he had never lent it much credence, Darwin himself had never entirely rejected the possibility that some acquired traits might be heritable, a fact that enabled those of his interpreters, in France and elsewhere, who wished to read him as continuing in the Lamarckian tradition. But Weismann's discovery, while compatible with Darwinian evolution by variation and selective retention, was manifestly *incompatible* with Lamarckian evolution by the inheritance of acquired traits. The neo-Darwinism that followed the incorporation of Weismannianism thus posed significant challenges to the conceptual framework under development since the 1870s, on which the nation's evolutionary destiny might be secured by education and other cultural interventions.

But in spite of these challenges the synthetic imperative persisted, as it had since the beginning of the Argentine nation. The country's intellectuals continued to labor under the obligation to absorb the latest the "civilized" world had to offer, select what was most useful for the modernization of the country, and perform the synthesis necessary to lend the results of this selection some semblance of coherence. In pursuit of synthesis, intellectuals tended toward an almost encyclopedic eclecticism common throughout Latin America during this period, but especially pervasive in Argentina. Discussing his colleague Piñero, Krueger remarked to Wundt, "He holds a second appointment as Professor of Histology at the Faculty of Medicine, and is also the department head in a psychiatric clinic—the way all gifted men here develop an amazing number of specializations" (cited in Taiana 2005, 398).

As Argentina's brand of Spencerian social thought was on the wane, new syntheses were being undertaken, many of them far more congenial to a Wundt or Krueger.

Carlos O. Bunge and the Development of Social Psychology

Carlos Octavio Bunge, born into another prominent *porteño* family in 1875, began to publish in psychology at the end of the nineteenth century. The son of a Supreme Court Justice, he received the finest education Argentina had to offer, becoming steeped in the extraordinary synthesis of evolutionary thought that dominated the Argentine intellectual life of the era. The descendent of German-speaking immigrants, Bunge also read German and was powerfully influenced by Haeckel, Wundt, and Weismann.

In someone so thoroughly imbued with Argentine evolutionary thought, it is unsurprising to find early awareness of the profound crisis that afflicted evolutionary theory at the end of the century. In an 1894–1895 essay entitled "Los dominios de la psicología" (The Domains of Psychology),

Bunge argues that the great diversity of evolutionary conceptions proliferating at the time was evidence of the need for a unifying metaphysical project. This need was all the more urgent in the wake of the collapse of Spencerianism, which had furnished the philosophical grounds for believing in the unity of science. He brands the "rejection of 'metaphysical conceptions of the universe'" as an error "grounded in the contemporary reaction against scholasticism and rationalism." In itself, the reaction against a purely "deductive, scholastic, rationalistic, speculative method" is perfectly reasonable, and the fact that today's intellectual leaders, rather than merely "*make* the public *reason*, must *make* them *observe*" constitutes progress of a sort. But any "inductive, positive" exposition can serve this purpose, and by now "no one is prepared to accept a dogmatic synthesis." The time has come for psychology to return to metaphysics; "I believe in a positive metaphysics, or in other words, in transcendental psychology" (C. Bunge, 1927b, 12–14, emphasis in original).

Bunge's generation is "leaving the time of Comte and Darwin behind," and while the inductive method remains suited to expository tasks, it is not, itself, essential to scientific research. For "science remains today as it was yesterday: constructed on hypotheses. How, then, does the [inductive] method differ from others? Principally, in that it is *more hypocritical*, because it better disguises the origins of its doctrines and constructions, and in that it is *better informed*" (15, emphasis in original). In a Kantian vein, Bunge proposes that transcendental psychology, or positive metaphysics, begin with the careful determination of the boundary between the knowable and the unknowable. "In transcendental psychology," he asserts, "a nearly mathematical degree of exactitude is much more attainable than in psychophysiology or rational psychology.... To distinguish the knowable from the unknowable is to distinguish, within the human psyche, the concrete from the abstract. With regard to the concrete, exactitude is only ever relative; but with regard to the abstract, it is possible to conceive it in absolute terms" (16–17). Transcendental psychology is thus an abstract, a priori discipline. It is a necessary complement to the concrete discipline of psychophysiology, not its replacement.

During that same period in the middle 1890s, Bunge also sketched a contribution to psychophysiology, his theory of the "unconscious-will" (*inconciente-voluntad*). From Haeckel, he derived the notion of "memory conceived as a general function of organized matter." Thus understood, much, if not most, human memory might be unconscious, and what intrigued Bunge was the prospect that the ability to think rested on "*the function of unconscious memory, whose activity is infinitely more valuable than that of conscious memory"* (C. Bunge 1927d, 50–51; emphasis in original). Where Haeckel provided the biological framework for the existence of the unconscious or subconscious (which Bunge would later distinguish), Wundt's work on hypnosis and suggestion supported the same conclusion (Wundt 1893).[11] Following Wundt, Bunge affirms the existence of *"regions of the subconscious not normally in evidence"* (C. Bunge 1927d, 65). In Bunge's view, the existence of a dynamic, independent subconscious was indubitable, and as for those who denied it, "pride will not allow them to see that their own minds contain an obscure, vast, active, and powerful back room," a place where "perceptions, sensations, and images live in constant motion, unseen, like the subterranean labors of legendary gnomes" (48).[12]

The role played by the principle of subconscious-will in Bunge's work is similar to Freud's later notion of the dynamic subconscious.[13] And although Bunge never developed the analysis of dreams as a means of understanding the unconscious, it is clear he was aware of this possibility. In a section on dreams and somnambulism, Bunge asserts that when faced with a preoccupation, one may "wait until a series of subconscious operations related to the dominant preoccupation occurs in a dream. On awakening, these operations may illuminate the concern, and have conscious consequences" (C. Bunge1927d, 64). But the primary thrust of Bunge's theory was the problem of heredity, the same problem that confronted the far more senior José Ramos Mejía during this period. According to Bunge's theory of the subconscious-will, *"the hereditary transmission of psychic predispositions"* was possible, predispositions that could come to *"constitute real emotional states, and even virtual or latent ideas"* (66; emphasis in original).

In publishing this theory, Bunge's intention was to counteract the extensive influence of Ribot, so evident in Piñero. For Ribot, "consciously directed willpower was the last of the mental qualities appropriated phylogenetically in mankind's slow evolution from savagery" and hence a faculty inaccessible to members of the less evolved races. Clearly its absence would undermine masculinity, because the superior man was distinguished by his commanding will (Nye 1984, 128). Bunge thus makes the possession of consciously directed willpower independent of heredity. For Bunge, this willpower isn't conscious, but rather unconscious, "a force x, whose essence is not cognizable."[14] He calls it the *"law of instinct,"* which "might with equal justification be called the *law of life*" (C. Bunge1927d, 80–81).

His proposal, like that put forth at the same time in Henri Bergson's *Introduction à la métaphysique*, emphasizes the role of instinct as a psychic force essential to life (see Bergson 1913).

In his attempt to create a psychological system founded on a "law of life" in place of the evolutionary "law of death," Bunge offers Argentine scientific culture a ray of hope. Darwin need not be discarded, for "Darwinism never denies the possibility of an ideal psychic principle," but Spencerian evolutionary theory, on which everything is based on "the mechanical transformation of material forces" (C. Bunge 1927d, 91), may be ruled out of court. For Bunge, whose Kantian predilections we have noted, the instinct-will[15] is understood in almost noumenal terms, as an inexplicable part of "the mystery of the unknowable" (C. Bunge 1927d, 94).[16]

Bunge's "Notas de psicología social" also appeared in 1895.[17] In this study, Bunge goes beyond the Spencerian view of society as a mere organism, calling it instead "*a psychic organism*" structured around humans, defined as "*an animal who aspires toward his own improvement and perfection*" (C. Bunge 1927c, 126; emphasis in original). He applies his theory of aspiration toward solving the problems of race that plagued Argentina, for after all, it was quite possible that there were no men anywhere who more aspired to being "civilized" than Bunge's Argentine compatriots. It is impossible, he claims, echoing Nietzsche on aspiration, "to conceive that a race might march in the vanguard of civilization if it doesn't know how to aspire." The black population of the United States, for example, which in Bunge's view had riches within its grasp, had remained stationary because "their aspirational potential is significantly less than that of the whites" (129). The same did not hold, however, for Argentina, where the great progress that had been achieved was owed to the desire for entry into the world of nearly perfect beings known collectively as civilization.

In 1903[18] *Principes de psychologie individuelle et socielle* (Principles of Individual and Social Psychology) appeared, a collection of papers on social psychology (C. Bunge 1903c),[19] including one on Bunge's theory of the instincts, a topic that was already the subject of much discussion in psychology. For Bunge, however, the study of the instinctual life was connected with his response to the aforementioned synthetic imperative: the search for a principle that reunited the different strands that made up the fabric of Argentine intellectual life. Auguste Dietrich, a leader in the fields of psychology and race, translated the book and wrote a preface in which the ambition and peripheral nature of Argentine intellectuals is clearly recognized. In his view, the "Argentine ambition [was] to exercise a sort of spiritual mastery, one sometimes prone to a bit of hyperbole, over every order of idea." But Europe was "not yet prepared to receive such revelation or such protection from America, be it from the northern or southern latitudes. We have, as yet, no need of rescue by this new, unforeseen Good Samaritan." He is also quick to claim that such intellectual pursuit was related to the "preponderance of French influence" that had allowed Argentina to "almost entirely shed its Castillian skeleton" (1–2).

Bunge's ambition was, in fact, to save European thought from the philosophical crisis brought by on Darwinian evolutionism. In 1904 he explains that, for him, a theory of instinct serves to reconcile "the facts established by intellectualist metaphysics (Kant, Hebart, Fichte, Schelling), voluntarist metaphysics (Schopenhauer), and biology (Darwin, Haeckel, Wundt)." The exposition of any such theory would have to begin with the acknowledgment that evolutionary thought had thus far failed "to explain the connection between internal experience and external experimentation" (1927e, 36) and set out to repair this deficit, which Bunge views as a consequence of the limitations of materialism. "Though all the psychological systems engendered by cosmological conceptions are highly imperfect, only materialism has closed off the way to future progress. This failure is a consequence of its irreparable error with regard to the theory of knowledge, an error materialism commits from the very first instant, in the very foundation of its construction. It ignores or discounts three fundamental facts: 1. Internal experience has priority over external; 2. Objects in the external world are, for us, images or representations (*Vorstellungen*), engendered in accordance with psychological laws; 3. The concept of matter is entirely hypothetical" (39; German term in original text). In a substantive departure from the "material monism" (40) with which earlier Darwinism was so closely associated, Bunge emphasizes the irreducibility of subjectivity. In this he was profoundly influenced by Wundt's philosophical psychology. Within the system he proposes, unity is to be sought in "the unity of human thought, derived not from any necessary objective unity of Nature, but from the subjective, organic unity of man" (40). This system is grounded on two "fundamental postulates," one for physiology (understood as the product of biological evolution), and one for (philosophical) psychology: "1. For physiology, the existence of a psychic nexus within seemingly automatic nervous phenomena; 2. For psychology, the unity of the psyche, via the extension and significance of the subconscious-subwill [*subconciencia-subvolundad*] and the determinacy of the conscious-will" (35).

Bunge was addressing the central synthetic problem of Spanish American philosophy during this period, the problem of reconciling materialism with idealism. The necessary unifying principles, in his view, were to be sought in the complementarity of philosophical and physiological psychology, where the former derived from Kantian transcendental psychology, and the latter from Darwin's observations. In forging this synthesis, he self-consciously eliminates all trace of Spencerian metaphysics.

> The metaphysical conception that emanates from Darwin is not unacceptable; the same, however, may not be said of Spencer's concept of the world, insofar as he exaggerates the law of evolution and ascribes a purely mechanical origin to psychic phenomena. Without a doubt, in their prudence and parsimony, the eclectic doctrines of Wundt appear far more scientific at the present time. The most important thing is not to assert, as incontrovertible truths, any hypotheses that truck with the unknowable. Applying these principles to the theory of instinct, with its positive, experimental character, I find myself conceiving it neither in materialist nor in idealist terms, though I am more inclined toward the latter. The future will tell which of these two hypotheses is more satisfactory. It may be that a third has yet to be discovered or invented, one that resolves the antinomy of materialism and idealism by means of a common principle, still unknown or badly interpreted. It might even be instinctism. (1927e, 43)

Bunge's quest for a transcendental psychology was compatible with his growing interest in social psychology. *Nuestra América* (Our America), an attempt to analyze Spanish American reality in psychological terms, appeared in 1903. The book opens with a preface aimed at explaining not only the author's country, but his education and the personal problems that prompted him to write. Bunge relates that he was raised with a belief in the existence of an absolute idea of the Good, a notion that he, "by psychological inheritance, by temperament," found easy to grasp. Throughout his childhood, his soul was divided into three parts, "the Good, the World," and the "Ego" (Yo). As regards the latter, he believed "in my Ego, in the *aspirative potential* of my individual" (C. Bunge 1903b, 2). But he soon discovered the imperfections of humans and plunged into a terrible depression that nearly killed him.[20] His fear of death saved him, showing him that "to live it was necessary to rebuild" the three parts of his soul, which had been destroyed. The World, he constructed around the concept of the aspiring man, very much inspired by the work of Friederich Nietzsche, as influential in Argentina as elsewhere in Spanish America. The Good would now be determined by the ideas of Happiness and Progress. Finally, his Ego would be reconstituted by the study of his native country, a country he now meant to lay out "like a cadaver on a dissection table," to "sort its tissues with a scalpel" (C. Bunge 1903b, 13).

In yet another 1903 work, an essay entitled "Aristocratizarse: Una concepción sintética de la historia" (On Becoming Aristocratic: A Synthetic Conception of History), Bunge returns to the philosophical issues of the day, this time attempting a synthesis grounded in a biological interpretation of history. After briefly explicating an evolutionary species concept, he asserts, "Biology teaches that the higher one rises on the animal scale, the easier *degeneration* becomes. Nothing is thus more prone to degeneration than man" (C. Bunge 1903a, 330). According to this view, historical conflicts become expressions of degeneration and the struggle against degeneration. Borrowing aspects of Marxism, then gaining currency among young intellectuals, Bunge reasons that successive historical conflicts ought to be associated with successive ideological movements. But in that case, he asks, "If the individualist philosophy of the French Revolution was romantic, and if its legitimate successor, socialist philosophy, is metaphysical, then where do we find a truly *scientific* philosophy: one that, rather than contradicting the laws and truths of science, accommodates itself to them? Perhaps it has not yet been articulated?" (349–50). If not, the task of contemporary philosophy is clear. "More or less vaguely, Socrates, Plato, Hobbes, Bentham, Darwin, Comte, Mill, Spencer, and above all Nietzsche, all laid the groundwork for a scientific philosophy. Socrates and Plato were its prophetic visionaries; Hobbes, Bentham, Comte, and Mill, its mere predecessors; Darwin, the formulator of its scientific foundations; Spencer, in the few valid portions of his work, a brave applier of Darwin's. . . . It is a great shame that Nietzsche, who was born in a metaphysical country and expounded his intuitions metaphysically, should have died before constructing a complete system" (353). Echoing aspects of Nietzsche's account of the "Overman," Bunge concludes that the perfection of the human species must be sought "*in relation* to one's conspecifics." It consists in becoming superior to other humans, whether particular individuals or groups. "Here, then, we find the general formula: *to progress is to become aristocratic*" (353; emphasis in original). In Argentina, as in Europe, both the left and the right drew inspiration from evolutionary theory. The incompleteness of its constitutive analogies made them available to this open-endedness and gave impetus to those who sought, with Bunge, to ground them in a synthetic philosophical system. Like Bunge, José Ingenieros turned to psychology in search of the necessary synthetic principles.

José Ingenieros and Evolutionary Psychology

José Ingenieros was born in Italy in 1877 but educated in Argentina (Bagú 1936). He received his medical degree in 1900 under the supervision of José Ramos Mejía, with a thesis entitled *Simulación de la locura* (The Simulation of Madness). He participated in the subsequent political reorganization and was active as a writer, physician, sociologist, and politician, pursuing the cultural multiplicity typical of his times. While still a student, he founded *La Montaña*, along with writer Leopoldo Lugones, to promote socialist ideas, but he changed his views, coming to understand history not as a class struggle, but as a race struggle (Ingenieros 1913, 41–42). The influences on which he drew were varied: "Darwin, Spencer, and Le Dantec in biology, Lombroso, Ferri, Charcot, Maudsley, and Morselli in psychiatry and criminology, and Ribot, Sergi, Mantegazza, Paulhan, and Nordau in psychology. He enjoyed direct contact with some of these figures, like Ribot and Le Dantec in Paris, and Ferri and Morselli in Rome" (Loudet 1971, 18–19).[21] His differences with Bunge are apparent from his 1903 critique of *Nuestra América*, published earlier that year. "Bunge concludes the first book of his study by asking whether the disease [of race] is incurable. He finds only one treatment: we must 'Europeanize' ourselves by 'work.' There is no attempt to confront the possibility that Europeanization might violate our character; 'indolence doesn't build character, it undermines it.' 'If the nature of Spanish American character is the lack of character, we will have to make ourselves a new character'" (Ingenieros 1903, 698). What bothers Ingenieros is Bunge's "unscientific illusion" of free will, coupled with a psychology whose central principle was the avoidance of pain and the pursuit of pleasure. He is frankly skeptical "that South Americans can modify their character by simply willing, or, if they lack character, invent one." "'Europeanization' isn't a goal, as Bunge claims, it is an inevitable event, one which would take place even if Bunge and the whole of Spanish America opposed it" (703).[22] Unlike Bunge, Ingenieros's interest in human volition is strictly connected to biological processes. Even when he admits the role of the will, it operates only within material developments. For this reason he is completely unambiguous on the issue of miscegenation, asserting that "the Argentine experience has shown the nefarious effects of hybridization" on the population.[23] For Ingenieros, "the races of humanity are fundamentally different, unequal, and unequally civilizable." Human equality, for him, was no more than "the noble dream of such innocents as Christ and Bakunin" (Ingenieros 1908b, 277).

The essay "La simulación en la lucha por la vida" (Simulation in the Struggle for Life) appeared in 1903 as the introduction to the published version of his doctoral thesis, after its main ideas had been adumbrated in various lectures and papers (see, e.g., Ingenieros 1901). Surprisingly, we find Andrew Lakoff placing this work in the context of the neo-Lamarckian ideas that continued to dominate much of the social sciences (Lakoff 2005). Lakoff barely mentions Darwin's influence, though Ingenieros's text makes it abundantly clear. This omission is hardly unique; it is representative of the incompleteness of much analysis of Argentine evolutionary thought. As we have noted, at the end of the nineteenth century, neo-Lamarckian ideas on soft inheritance were, indeed, very influential, usually in their Spencerian variants. But by 1903, the psychiatric community to which Ingenieros belonged had generally come to accept that these were biologically ill-founded. Argentine evolutionists were struggling to come to grips with the debates between neo-Lamarckians and neo-Darwinians that had cast evolutionary thought into disarray. It is for this reason that Ingenieros drew on ideas from different sources in an effort to forge a coherent school of thought on issues that were so crucial to national evolution.

Ingenieros's interest in simulation represents his attempt to shed light on the psychological mechanisms that might be brought to bear in this struggle. His interest in specifically *psychological* mechanisms doubtless helps explain his choice of terminology, and in particular his use of *simulación* (simulation) in place of *mimetismo* (mimicry). *Mimicry* is the word used in the well-known discussions of Henry Bates (1825–1892) and Fritz Müller (1821–1897), with which Ingenieros was surely familiar.[24] *Simulation* and its Spanish cognate carry a more voluntaristic connotation than *mimicry*, suggesting that Ingenieros had opted to psychologize a natural mechanism, rather than naturalizing a psychological one. Indeed, the very first example of "simulation" cited in "La simulación en la lucha por la vida" is that of a camouflaged larva. Ingenieros does not attribute deliberative rationality to the larva, but he *does* attribute intentionality. The larva is an active participant in its struggle for life: an agent. Such agency is strictly biological, but it is also a primitive psychological attribute; in phylogenetic terms, it is *the* primitive psychological attribute. As an adaptive mechanism, the simulation to which an agent in the struggle for life may avail itself is found throughout the animal kingdom; it consists in the activity of an individual organism bent toward preserving its life by pretending to be something else. At bottom, it is a biological phenomenon, not a social one—but once this has been acknowledged, we may proceed to explore "its fully developed conscious manifestations in superorganic life—in human societies" (Ingenieros 1917, 110). The foundation of this exploration is clearly Darwin, who provides the constitutive analogies for his understanding of biological evolution; Ingenieros extends them to social structures. This extension, Ingenieros announces, is made possible "by the most recent inductions of sociology, and by employing the most rigorous scientific method" (111).

In other words, Ingenieros's thesis regarding the social phenomenon of simulation is based on the "Darwinian doctrine . . . the premise that sustains the entire development of this essay." This doctrine consists in the "variability of species, or transformism, [which has] been confirmed by all the biological sciences, without the objections raised by its adversaries on certain matters of detail having significantly affected it" (Ingenieros 1917, 114). Ingenieros further describes the struggle for life in nearly the same terms as Darwin himself, as "the doctrine of Malthus applied, in all its intensity, to the beings of the animal and vegetable kingdoms," adding, as Darwin had, that "the phrase 'struggle for existence' is used in a general, metaphorical sense" (115).

Next Ingenieros turns to the troublesome question of inheritance, with the disclaimer that a full discussion of current debates on this issue would go beyond the scope of his study. Instead, he refers readers to the works of "Lamarck and Darwin, Kolliker, Wagner, Naegli, Weismann, Mantegazza" and other advocates of "the modern neo-Lamarckian and neo-Darwinian schools." It is important to note that Ingenieros feels no compulsion to take sides, here. He takes the debate to be an internal discussion among biologists. Given the constant rush of new discoveries, it is important to leave room for such uncertainties. "The struggle for existence in human societies is an undisputed fact that manifests itself in ways similar to those observed in the biological realm; and this truth is equally admissible to those who believe in Spencer's biosociological doctrine, for whom human societies are simply superorganisms, as it is for those who accept the primacy of economic phenomena in the constitution of societies, with or without the theory of class struggle, one of the foundations of the poorly named 'historical materialism'" (1917, 118).

It is in his explication of the significance of the phenomenon of simulation that we find the greatest kinship between Ingenieros and Darwin. Ingenieros distinguishes imitation, which had been central to the work of Gabriel Tarde, from simulation proper, on the grounds that while the former implies that an act has been committed, the latter may constitute a mere appearance. This difference is important because of the emphasis placed within social psychology on contagion and imitation among the masses. But following Darwin, Ingenieros is interested, instead, in the psychological counterparts to biological processes found in all species. On this score, he cites Darwin's *Expression of the Emotions*, published in 1872, in which the English naturalist had observed,

> Even the simulation of an emotion tends to arouse it in our minds. Shakespeare, who from his wonderful knowledge of the human mind ought to be an excellent judge, says:
>
> "Is it not monstruous that this player here,
>   But in a fiction, in a dream of passion,
>   Could force his soul so to his own conceit,
>   That, from her working, all his visage wann'd;
>   Tears in his eyes, distraction in 's aspect,
>   A broken voice, and his whole function suiting
>   With forms to his conceit? And all for nothing!" *Hamlet*, act ii, sc. 2

(C. Darwin 2006, 1476)

Inspired by Darwin, Ingenieros also turns to Shakespeare—though to *Othello* rather than *Hamlet*. Darwin had been content to conclude that simulation "deserves still further attention, especially from any able physiologist" (2006, 1477). Following in Darwin's footsteps, Ingenieros takes this mechanism as essentially biological, but with psychological and social manifestations: we engage in simulation to further our chances in the struggle for existence, and this struggle is both biological and social.

Discussing the connection between instinctive and conscious simulation, Ingenieros reminds his readers that in "expounding the doctrine of the 'struggle for life,' we took this expression in a *figurative sense*, as Darwin did when he coined it." Similarly, we must understand simulation as a means in the struggle for life in a figurative sense, too. If we were speaking literally, then "we could only talk of struggle and simulation with respect to human phenomena, construed as *conscious* or *voluntary*" (Ingenieros 1917, 126; emphasis in original). In Ingenieros, as in Darwin, we find an acute awareness of analogy and metaphor; Darwin, Ingenieros makes clear, had drawn on the sphere of human action for his analogies, as Ingenieros now draws on Darwinian evolution. Over geological and historical time, the two spheres run together. "From his most primitive animal state to his present civilized condition, man has steadily abandoned violent means in the struggle for life, while steadily increasing his use of fraudulent means. Given the growing tendency of men to associate themselves against nature, it is likely that in future forms of local organization simulation will also decrease, in proportion to the attenuation of the struggle for life" (296).

Ingenieros is, perhaps, a stricter Darwinian than any other thinker canvassed in this chapter. But he, too, operated under the synthetic imperative. He asserted, with Spencer, the existence of social counterparts to biological mechanisms, and he followed Nietzsche's call for the pursuit of individual superiority. In addition, while as a Darwinian he understood that, given constant evolution by variation and selective retention, any talk of fixed types was ultimately unfounded, his psychological typology ultimately re-created racial typologies. Racial diseases became psychosocial manifestations of natural selection and of the futile resistance of complex, conscious beings against their evolutionary destiny.

Ingenieros's synthetic attempt received its most complete expression in his most well-known book, *Principios de psicología* (Principles of Psychology), which appeared in 1910. This is a pioneering work in the field of evolutionary psychology, one whose importance has largely been forgotten. It was published in 1910, in the context of the triumphalism of the centennial of Argentina's 1810 revolutionary movement. This context frames its totalizing attempt at a definitive theory of evolution. It went through six Spanish editions, one in French, published by the prestigious house of Alcan in 1914, and one in German, published in 1922 with an introduction by chemistry Nobel Laureate Wilhelm Ostwald (Triarhou and del Cerro 2006a).

In the various editions, Ingenieros continues to deal with the ongoing problems that contradictory European views had brought to his synthetic project. Like Rivarola, Quesada, and Bunge, he draws attention to the philosophical problems raised by Darwinism. But unlike Bunge, his solution is historistic (or "genetic"), not transcendental.

> The decisive factor in the [recent] general transformation of philosophy has been the extension of the idea of the incessant transformation of everything that exists, as confirmed within every domain of experience. This functional conception of evolution has given rise, in turn, to the necessity of a genetic

method. Vaguely intuited by thinkers of all epochs, this method was explicitly formulated in the nineteenth century, acquiring the precise contours that allow it to be applied to all manifestations of known reality.... Since Lamarck and Darwin, evolution has been applied to the emergence of psychic functions in the biological series ... preparing the way for genetic psychology. (Ingenieros 1919, 28–29)

Though Ingenieros greatly admires the scope of Spencer's ambitions, the heterogeneity of his principles had left "large gaps to be filled ... and firm conclusions resting on particular hypotheses whose inaccuracy has recently been demonstrated." Ingenieros affirms that "the unity of the real (monism) is incessantly transformed (evolutionism) by natural causes (determinism)." The "modern expression of [this] evolutionary philosophy" must rest on what he calls the "metaphysics of experience" (30). With regard to the "genetic" component of this scientific philosophy, Ingenieros recognizes that a great deal turns on the outcome of controversies still raging between neo-Lamarckians and neo-Darwinians. He holds out hope for a compromise, drawing heavily on the work of George Romanes and American psychologist James Mark Baldwin, who in 1896 had proposed that, under certain circumstances, Darwinian natural selection could lead traits acquired culturally, in one generation, to become heritable in subsequent generations (Baldwin 1896).[25] He credits Romanes with the invention of comparative, phylogenetic psychology, and Baldwin with rescuing Haeckel's biogenetic law and applying it to the evolution of psychological traits (Ingenieros 1919, 127). Drawing on Morgan and Osborn, Baldwin had demonstrated that "natural selection always operates on a combination of congenital characters and acquired variations, and not on congenital characters alone" (Ingenieros 1919, 106–7). But, according to Ingenieros, Baldwin had contradicted himself in denying "the heritability of psychological variations acquired by individuals." In his desire to maintain the transmission of such traits, Ingenieros insists on the importance of the social environment, because "every human being gains his experience in harmony with his social environment, within which he inherits the acquisitions of collective experience. This, and nothing else, is what we understand by 'psycho-social inheritance [*herencia psíquica social*]'" (Ingenieros 1919, 107). For Ingenieros, psychological functions are thus both part of the evolutionary legacy of our species and factors that contribute to our evolutionary destiny; "All the results of comparative psychology converge to demonstrate the mental descent of man as consistent with transformism or the doctrine of species variability" (Ingenieros 1919, 138–39).

Ingenieros's synthesis was the last important one before an antipositivist reaction directed by Alejandro Korn began to recalibrate the philosophical compass of the country. The chasm opened by Darwinian science was bridged by approaches less mechanicist and opened to more vitalist concerns that fit better with the intellectual life of Argentina. Science would continue to be important in the country, but its limits in the process of nation building would become undeniable. And while Darwinism would remain significant, it would no longer be the dominant intellectual current in national culture.

Conclusion

Tomorrow we will open the window again, hoping to attain, detail by detail and to the best of our ability, some clearer notion of the whole. — Rodolfo Rivarola, 1896

As we have argued above, one of the most notable characteristics of evolutionary thinking in Argentina was its synthetic imperative. By the 1920s, as Schaub noted, Argentine philosophical thinking had developed its own national characteristics. Argentine Darwinism grew in response to the needs of nation building that characterized the second half of the nineteenth century. In following closely the universalist mandate that emanated from Enlightenment thought, Argentine intellectuals attempted to integrate all civilized ideas into the country's civilizatory process. The Darwinian intellectuals discussed in this book claimed to address problems both local and universal.

In the case of Argentina, the need for renewal and unification responded to the same foundational principle that had organized the struggle against Juan Manuel de Rosas in the name of civilization by the mid-nineteenth century. The universalism that provided the philosophical grounding of the nation can be observed in the work of Estéban Echeverría, a leading defendant of European ideas who died while exiled in 1851. In 1848 he reflected on the revolution that had recently given rise to the Second French Republic, with particular attention to the ideals of unity and fraternity. Behind the political transformation, there lurked a philosophical synthesis that rested, in part, on pre-Darwinian notions of evolution. "In order to be stable and fruitful, progress must be normalized. It must be accommodated to a law of successive, incremental development, and this law, in turn, must be embodied in the historical and educational traditions of the society" (E. Echeverría 1873, 437). Such progressive, social evolution, of which the

Revolution of 1848 constituted an example, is the "manifestation of the synthetic thought which philosophy has instilled in the breast of French society, and which society, in turn, has gradually refined." In France, as elsewhere in Europe, as Echeverría understood it, progress was a function of reason, expressed in civilized thought. Following Saint-Simon and Leroux, Echeverría emphasizes the role of philosophy as custodian of the all-important *unity* of such thought, which allowed it to encompass the full range of change, from the spiritual to the biological, conceivable within a civilized society. "This is not abstract thought, a mere isolated part of reason, but rather a rational conception deduced from knowledge of history and of the animated organism of society, elevated to the status of a law governing the successive emergence of the social phenomena that constitute the life of a nation" (440). In the succession of these phenomena, the "communion of all men" announced the beginning of an era of limitless perfectibility. History was simply the "successive education of humanity." The function of philosophy, or synthetic thought, was to discover, in history, the "regenerative principle" whereby society evolved from a mere "agglomeration of human beings divided into castes, perpetually hostile to one another" into a "communion of solitary creatures" capable of working "for the reciprocal good and perfection of one another." Under the tutelage of philosophy, humanity would come to constitute "a single family united by the bond of this same law, and the Holy Alliance of Peoples prophesied by the French Revolution in 1792 will come to pass, putting an end to all servitude and tyranny" (454).

Implicit in Echeverría's vision, as in the work of others of his generation, is the notion that the biological perfectibility of the human species, as guided by philosophy, and its future unity, will result from blending inheritance. As we have discussed, the irruption of Darwinism disrupted the implementation of this project. In the first place it exacerbated the different national conceptions of civilization and progress and the separation between French, German, and English philosophical traditions and their relationships with science. Second, the possibility of reconciling these different approaches to evolution required a metaphysical component: it could either occur within the confines of a materialistic determinism, or else Darwinism would have to be made to fit with a dualistic or idealistic program. Haeckel, Darwin (in the *Descent of Man*), and Spencer all contributed to the broadly materialistic (or monistic) efforts at synthesis that took place in Argentina. But by the turn of the century, these had run their course.

The next round of syntheses, influenced by such figures as Wundt, tended more toward idealism—or at least toward a deflationary, Kantian metaphysics. An article published in 1902 in the *Revista Jurídica y de Ciencias Sociales* (Journal of Law and Social Sciences) is an example of how Argentine intellectuals perceived this problem. Its author, Víctor R. Pesenti, explains his rejection of the views of Darwin and Lester F. Ward and his preference for a Comtian approach, in terms of the need for more generalizing principles to sustain the philosophical grounding of discussions of evolution and nature.

> We disagree with this author [Ward] and Darwin in their assertions that "there is no finality in nature," that "some organisms are more complex simply because more force has accumulated within them," and that "they contain only the concentration and localization of cosmic energy." Nonetheless, as Ward himself acknowledges, within certain organisms we observe "the emergence of a system of cooperation" entirely analogous to "the social system," which in the end is nothing more than a modification of the uniform, universal process of the concentration of cosmic energy, in turn identical to the force described by Spencer.... Science must be systematic, or become a poor rhapsody, incapable of satisfying the spirit of man; and in order to be systematic, it is necessary that the Universe be a system oriented toward certain ends. (Pesenti 1902, 23)[1]

Lester F. Ward was an influential figure at this time for his attempt to provide, through sociology, a systematization of post-Darwinian philosophical ideas and his defense of a Lamarckian view of inheritance (Rafferty 2003). He retained some principles from evolutionary biology—his "biological imperative"—but also proposed a teleological system to explain the evolution of human societies and human equality (Degler 1991, 351). It is for this reason that Pesenti attacks him for his defense of natural selection but at the same time praises his use of analogies to restore a privileged role for humans.

In Argentina, the crisis of Spencerian philosophy reaffirmed the need to renew the faith in a universal civilizatory principle that would assure the future of the nation. The destiny of sociology occupied the minds of many intellectuals by the beginning of the twentieth century. Most of them insisted on the need to reconcile pre- and post-Darwinian thought, despite mounting evidence in support of the opposing neo-Darwinian trajectory. It is important to remember that this process was not exclusive to Argentina; we may observe this interest in integrating evolution with universalism and cooperation throughout the Spanish-speaking world. Spanish lawyer Adolfo Posada, an influential positivist thinker, wrote in 1902 that the task of sociology was, among other things, "the application of Darwinism, with consequences even Darwin did not foresee, and perhaps would not even have accepted, toward the explanation of social phenomena" (Posada 1902, 232). Returning to the importance of analogies from the natural to the social world, Posada also insists on the relationship between analogical reasoning and culture, and the need to correct a type of Darwinism that served particular cultural interests.

> If Darwin's hypotheses had been interpreted differently, less naturalistically and cruelly, and above all, if the attention of the sociologist had been less toward the meaning, in nature, of struggle, combat, and violence, and more toward the infinite phenomena of cooperation and sacrifice, that in turn reveal the existence of a higher principle of expansive sympathy, then who knows? ... Perhaps instead of such blinkered scientific attention to struggle, leading as it does to the view of *man as the enemy of man*, we might find, instead, due attention paid to love, affection, mutual aid, sacrifice, and in short, everything that tends to unite souls and calm spirits. (233, emphasis in original)

The theory propounded by Darwin in *On the Origin of Species* was different in many respects from many other scientific theories that enjoyed currency at the time. One of these was its irreducible reliance on a particular way to use analogies. As Darwin acknowledged in his *Autobiography*, almost alone among the works assigned to him as an undergraduate of Cambridge, those of the late eighteenth- and early nineteenth-century natural theologian William Paley struck a chord in him (see Krasner 1990; A. Levine 2009). Paley explained the origins and diversity of organic life by recourse to an argument from analogy. Darwin, when he tried his hand at this explanatory task, invoked a somewhat *different* analogy; but his reasoning remained similar to Paley's, so much so that, as we noted in chapter 3, some of his most influential contemporaries in philosophy dismissed his theory as beyond the pale of the inductive sciences. It is here that our notion of science-consitutive analogy is most helpful. Darwin's theory of evolution, arguably the most important scientific discovery of the nineteenth century, is built from the ground up on analogies. In addition to the analogies, it also boasts theoretical hypotheses and numerous observations, but without the analogies, there would be no connection between the two.

The analogies of *Origin* and *Descent*—between breeds and species, between variation under domestication and variation in nature, between selection under domestication and selection in nature, between the Malthusian struggle for resources and the struggle for existence, and so on—are culturally contingent. While pre-Darwinian French and Argentine intellectuals emphasized the universality of the effects produced by an evolutionary process directed toward perfection and progress, Darwinism resisted compromise with this kind of universalism. The process might be the same throughout nature, but its consequences were contingent and variable. This approach made his use of analogies particularly sensitive to the environments in which his ideas were applied. Transported to the South Atlantic, they encountered and interacted with different cultural contingencies, including the budding liberal national ideology discussed above. The cultural possibilities of their English origins affected their scientific trajectory, as well as their broader social and political significance, and so, likewise, did the cultural contingencies of their Argentine reception.

The limitations of the constitutive analogies of evolutionary theory would become still more apparent in the early twentieth century, following the rediscovery of Mendel's work. The legacy of Haeckel, who lived until 1919, was also undergoing transformation. In 1900 José Ingenieros published a review of a French translation, published that same year, of Haeckel's 1898 pamphlet *Über unsere gegenwärtige Kenntniss vom Ursprung des Menschen* (On Our Present State of Knowledge Concerning the Origins of Man), originally an address to the Fourth International Zoology Conference in Cambridge (Haeckel 1898b; Ingenieros 1900). Ingenieros credits Haeckel with having made the emergence of experimental psychology possible by "situating the functions of psychic life within the nervous system," thus destroying "the entire armamentum of metaphysical psychology, including the belief that there exists in man a 'soul,' understood as an immortal, extraorganic entity." In its place, Ingenieros extols a Haeckelian monistic philosophy that, "founded on evolution, would greatly further the knowledge of natural truths, while its practical application would also enhance the pursuit of the Good and Beautiful" (Ingenieros 1900, 173). The essential purpose of this philosophy would be the "study of the origin of man" (171). In his defense of Haeckel, Ingenieros acknowledges that the future course of science is uncertain; but evolutionary theory, he claims, will remain, a fact that philosophy must acknowledge. More troubling still was the fact that some of the great scientific discoveries of the second half of the nineteenth century entailed that the principles governing important events, including biological processes, tended more toward contingency than necessity, and more toward particularity than universality.

In a 1904 article on science instruction in secondary education, Angel Gallardo concedes as much, noting that the goal of education was to instill, "in addition to enthusiasm for natural wonders, a notion of the role of man in nature, and a sense of the limitations and contingency of human science" (Gallardo 1904, 44). Such contingency fell out of the Darwinian principle of evolution by variation and selective retention, a poor fit with the Enlightenment proclivity toward universality and harmony. So long as this or similar principles were included in the mix, the challenges faced by any prospective synthesis were bound to be severe.[2]

In his 1907 monograph on Herbert Spencer (see chap. 5), Ernesto Quesada clearly explains the dilemmas faced by Argentine evolutionary thinkers. By this time the Spencerian synthesis had already begun to unravel, as it became increasingly clear that Spencer's work lacked scientific support. Why, Quesada asks, was the thought of an entire generation founded on a doctrine so plainly at odds with the best available scientific results? His answer is simple: to meet the challenge that Darwinism had posed to philosophy, it was necessary that its analogies be construed in such a way as to maintain the union of the social with the biological. It was inevitable that this necessity make itself felt in a nation that found itself, like Argentina, committed to establishing itself on scientific principles, so as to join the civilized world. Darwinism had disrupted the teleology of this national ideology. An analogical link between the biological and the social was essential to its reconstitution, and Spencer had provided one, simultaneously bridging the gap between pre-Darwinian and post-Darwinian thought.

Spencer also represented the continuity of certain pre-Darwinian notions, championed by Echeverría and persisting late into the century in the work of Rawson and Sarmiento. Spencer's Lamarckianism was particularly significant, in that it restored faith in the potential of superior culture to bring about individual improvements. Argentine evolutionary thinkers read Spencer in a manner not unlike Peter Bowler, who has remarked that he *"was really a Social Lamarckian.* He justified a society based on struggle not by reference to sexual selection, but by portraying struggle as the chief force encouraging self-development" (Bowler 1994, 113). Spencer's revival of Lamarckianism, together with his strong analogy between organism and society, resonated with the views of members of the generation of Argentine intellectuals born around Independence, for the continuity of the tradition of civilized thought was essential to the future development of Argentina. To paraphrase Quesada, Spencer *contextualized* Darwin. As Quesada understood him, Spencer's project allowed for the marriage of the newest scientific discoveries with both the pre-Darwinian intellectual legacy and a modernizing political project.

> Philosophy's lack—of a concept of the whole, a system of unification, a doctrine meant to serve as the backbone of the *corpus* of human knowledge—was not filled by Comte's positivism, despite the phalanx of English thinkers who embraced his credo. The original sin of this system, the incongruence between the positivism of the *Cours* and the dogmatic religiosity of the *Système*, was enough for the essentially logical mentality of the British spirit to resist it.... In short, we still wanted a philosophy capable of imposing itself upon the public, and which, while remaining strictly logical and scientific in its premises, might nonetheless aspire to the universality of knowledge, toward the renewal of the Baconian project, making the study of society and its diverse phenomena one of the many disciplines of the whole. (E. Quesada 1907, 19)

Alas, as Quesada acknowledges, this effort had failed. "The intellectual revolution initiated by Darwin" was simply "the beginning of a new era," with too much left to be discovered, including the laws governing "variability ... [and] inheritance itself" (23). Still, the burgeoning relationship between the natural sciences and psychology pointed the way toward new solutions to the metaphysical problems brought on by the substance and method of Darwinism. By the beginning of the twentieth century, it was clear that the lasting legacy of evolutionary theory would include the deployment of Darwinian analogies not only in biology, but also in political and social thought. As Rodolfo Rivarola noted in a 1908 piece on university reform,

> In the philosophical and scientific realm, we have borne and continue to bear witness to the influence of one dogma propounded as a law of evolution, and another associated with positivism and its methods. It is difficult to confine such influences to within a given sphere without extending them, by analogy, to one or another separate region of facts and phenomena. Thus the concept of evolution, as it applies to organic matter, may be easily translated to the realm of social thought and action; the concept of an organism may be extended to encompass societies, until the analogy or purely symbolic relationship is taken as real, to the point where no one doubts that society really is an organism. (Rivarola 1908, 215)

If the specifically Spencerian analogy had been undermined, hope that the synthetic potential of psychology (and by extension, the social sciences) might furnish a new philosophical foundation for the broader application of Darwinian analogies persisted. 1908 also saw the creation of the Sociedad de Psicología de Buenos Aires (Buenos Aires Society for Psychology), the second such society in Latin America, after its Mexican counterpart. The guiding lights of Argentine evolutionary thought were all among its founding members, including Florentino Ameghino, Carlos Octavio Bunge, José Ingenieros, Alejandro Korn, Victor Mercante, Horacio Piñero, José María Ramos Mejía, Rodolfo Rivarola, José Seprun, and Rodolfo Senet (see Sanchez Sosa and Valderrama-Iturbe 2001). As we discussed in chapter 6, the interest in psychology as a cornerstone of the scientific and social development of the nation dated back at least to the establishment of the Laboratory of Experimental Psychology in 1904 (see Argentina 1904, 671ff).

The mounting influence of German philosophical psychology in the early twentieth century had given further ammunition to those who sought a unified, metaphysical foundation for science and were convinced that materialism failed to provide such. On the occasion of his farewell address of 1908, Felix Krueger remarked on this trend. "Over the past decade, we have observed in all intellectual nations a progressive disengagement not only from vulgar, dogmatic materialism, but also from naturalistic unilateralism, which once characterized the majority of scientific psychologists. Contemporary psychology is ever more intimately and methodically intertwined with its fellow special sciences. At the same time, it has reaffirmed and strengthened its traditional ties with philosophy" (Krueger 1908, 85). Krueger's address concludes by noting that the philosophical current he represented, "philosophical idealism," had "found its echo in Argentina," giving him confidence that Argentine psychology would "progress, so long as its relationship with the German psychological and philosophical spirit is not neglected" (85). Krueger's prediction would come true, in that German idealism would remain important long after his departure, particularly among those who continued to follow the post-Darwinian synthetic imperative after the decline of the Spencerian program.

In a 1918 address to the Academía de Filosofía, José Ingenieros offered a famous set of predictions regarding the future of philosophy. Many of them are based on his assessment of the need for a reform of philosophical language in response to recent upheavals in metaphysics. In the context of the "philosophical crisis of the nineteenth century," spiritualism, which he calls "the exaltation of the affective-ethical in opposition to the logical-critical," had crept in. But rather than "overcoming the cycle of rationalism," it had "returned to the illegitimate sources preceding it within medieval theology." As an antidote to such tendencies, he proposes subordinating metaphysics to empirical hypothesis, thus paving the way for "a true logic of hypothetical, nonexperiential knowledge."

> The metaphysics of the future, now in the process of formation, will have certain necessary characteristics: universality, perfectibility, anti-dogmatism, and impersonality. It will consist in the subsumption of all forms of experience, for they all point toward nonexperiential problems. . . . In order to pose metaphysical problems accurately, the complete reformation of philosophical language is absolutely indispensable. The exactitude of every logical process is conditioned by the exactitude of the terms in which it is expressed; a perfect logic may never be expressed in imperfect terms, nor may perfect conclusions be inferred regarding the perfect relations among imperfect terms. . . . The various experimental sciences having disengaged from philosophy, the architectonic transmutation of the latter will continue into the future, giving rise to metaphysics as a unique genre, whose purpose will be to formulate hypotheses regarding that which exceeds the bounds of experience in all of the sciences. A harmonic system that attempts to explain the nonexperiential by recourse to the experiential, mediated by constantly renewed hypotheses founded, in turn, on perfectible laws, will never give rise to two discordant discourses of truths; it will realize the synthetic unity to which all legitimate metaphysics aspires. (Ingenieros 1919, 80)

Ingenieros's interest in a synthetic philosophy that remedied past failures opened up a line of inquiry that would put an end to the dominance of positivism. As Riseri Frondizi explains, the influence of Kant, Fichte, Schelling, and Hegel all combined was important "to support the fight against positivism." More importantly, Argentina "turned to the antipositivist movements in Europe, to Bergson, to Neo-Hegelianism, especially Croce and Gentile, to Neo-Kantianism (the Marburg School)." According to Frondizi, "Bergson was the European thinker that helped Latin-America more to overcome the positivistic stage" (Frondizi 1943, 181–82). Alejandro Korn, the physician who had once served as guardian of the ill-fated Damiana, became the new leader of Argentine intellectuals who tried to bridge the distance between biology and society with a philosophy that responded to the concerns of his time.

The fact that Korn did not begin to teach philosophy until he was forty-six years old serves as a sign of his personal transition, which in turn reflected the national one. Earlier in his career, as a psychiatrist, he had been involved in the scientific discussion that dominated the late nineteenth century. Now "he was highly critical of the attempts of Comte and Spencer and recognized his indebtedness to Kant, Schopenhauer, Bergson, Dilthey, and to certain writers in historical materialism" (Kilgore 1960, 80). As W. J. Kilgore has recognized, Korn's function in Argentine culture "was analogous to the contributions which William James was making in the United States and Bergson in France." Korn's main interest was to promote "the development of the sciences and of incorporating into philosophy those insights which the latest scientific investigations might provide. He also insisted that there were aspects of human experience which were explained inadequately in the deterministic approach of his day" (82).

The emergence of Korn's work, and his opposition to Ingenieros's biological proclivities, ushered in a new era in scientific thought. As should be clear from our discussion of his *Principios*, Ingenieros, too, was influenced by Kant. But this work would turn out to be the last attempt at synthesis that sought to vindicate biological determinism in the interpretation of Darwinian analogies. The next chapter in Argentine intellectual history would be a very different one. For all that, the constitutive analogies of Darwinism persisted as ineradicable elements of the national discourse, as, in some ways, they persist to this day. The duality of evolution and extinction discussed in chapter 4, for example, remains particularly significant. We have seen how the assimilation of evolutionary thought allowed Argentina's intellectual leaders to transform their extermination of the indigenous population into its simple *disappearance* in accordance with natural laws. It is no coincidence that the fate of victims of the dirty war of the 1970s is described in precisely the same terms. As we suggested in our introduction, science-constitutive analogies are *incomplete objects*: nodes in the complex, variable webs of meaning to which cultures give rise. As elements of scientific discourse, this incompleteness accounts for the potentiality of the scientific research programs they constitute. As elements of the broader discourse of which scientific discourse itself is but a node, their potentiality is still more open-ended.

The fact, easily acknowledged but seldom exploited, that science-constitutive analogies are culturally situated, is one of the attractions not only of the study of Argentine evolutionary thought, but of peripheral science in general. There are times when quintessentially *peripheral* insights—Ameghino's interest in extinction, the persistent attempts of Argentine thinkers to construct an evolutionary synthesis, and the early recognition of the metaphysical problems brought on by Darwin's theory—have larger significance to the *central* processes of science. It is our hope that historians and philosophers of science among our readers will come away from this book with further incentive to look toward the periphery. In the process, they may shed new light on what Peter Galison has called the "central metaphor" in both history and philosophy of science: the duality (and priority) of observation and theory (1988). Analogy may be constitutive of both.

Notes

Introduction

1. All sources originally published in languages other than English are quoted in our translation, except where noted.

2. Huxley's lectures clearly lay heavily on Sarmiento's mind in 1882, as witnessed by his extensive reference to them in the eulogy to Darwin he delivered that year (see p. 167). As always, the uses to which Sarmiento puts his sources are most charitably described as "selective." Huxley's lecture, "A Critical Examination of *On the Origin of Species*," contains one scant reference to paleontological discoveries in the Americas (Huxley 2004, 82).

3. "I must frankly confess that situated as I am without the benefit of any edifying society, and lacking training in the rigorous description of the natural world, I must beg the indulgence of the naturalists in the hope they will forgive my errors and imprecision. The announcement of this new species having struck me as of interest, and not having anyone to whom to delegate the technical details, I have felt obliged, despite myself, to submit to the critical scrutiny of the wise, baring myself to them in all my ignorance. My excuse . . . is necessity: it is she who has thus exposed me, and not the presumptuous delusion of adequacy" (Muñiz 1845).

4. Readers familiar with Morton's "American Golgotha" will doubtless be struck by the aptness of this comparison after reading chapter 4.

5. As evidence for the fin-de-siècle reputation of Moreno's collections, see Trübner's 1898 survey of reputable institutions of scientific research, *Minerva, Jahrbuch der gelehrten Welt*.

6. The "modern synthesis" in evolutionary theory consists in the early twentieth-century reconciliation of Darwinian natural selection with the theoretical framework of population genetics. It eradicated remaining traces of teleology from evolution; under the modern synthesis, as under Ameghino's earlier phylogeny, evolution involves *change* but not necessarily *improvement*.

7. See chaps. 4 and 6 for discussions of the importance of the concepts of degeneration and atavism in Argentine evolutionary thought.

8. See also Beatty 1985. David Stamos (1999) has argued against the interpretation of Darwin as a species nominalist.

9. Note that by *Darwinism* we do *not* mean social Darwinism. Social Darwinism, too, was an intellectual response to the social challenge posed by Darwin's discoveries, albeit a response by intellectuals (e.g., Herbert Spencer and William Graham Sumner) from societies quite different from those considered here (Britain and the United States, respectively). We use *Darwinism* as an umbrella term for the whole complex of views that emerged in Argentina in response to the Darwinian revolution. In addition to Darwin himself, this response was shaped by the contributions of an eclectic mix of European authors, including Spencer, as well as by the particular sociocultural context of nineteenth-century Argentina itself.

Chapter 1

1. For an account of the avid interest this volume provoked in Thomas Jefferson, see J. Boyd 1958. On this, see López Piñero and Glick 1993.

2. After breaking up with Spain in 1811, this area was disputed by both Argentina and Brazil. Uruguay declared its independence in 1825, a fact fully recognized by a treaty signed with its neighbors in 1828.

3. Darwin Correspondence Project, Letter 215: Darwin, C. R. to Darwin, C. S., 20 Sept [1833]. Except where otherwise noted, letters to and from Charles Darwin are cited as they appear online in the Darwin Correspondence Project, www.darwinproject.ac.uk.

4. According to William Katra, the members of this generation "were the sons of the burghers who led Argentina's 1810 revolution against Spanish rule that erupted after the Napoleonic invasion of the Iberian Peninsula. . . . As children they witnessed and suffered the decades-long civil strife that the independence struggle had ignited. Beginning in the 1830s and continuing for more than five decades, they sought to transform liberal ideals into veritable institutions and practices through their writings and sociopolitical activism" (1996, 7).

5. This shows the close attention that de Angelis paid to anything published in Europe that might be relevant to Argentina. In fact Darwin's *Journal of Researches* was first published in 1839 as the third volume of King and Fitz Roy's *Narrative of the Surveying Voyages of Her Majesty's Ships Adventure and Beagle between the Years 1826 and 1836. Describing Their Examination of the Southern Shores of South America and the Beagle's Circumnavigation of the Globe.* Vol. 3, *Journal and Remarks, 1832–1836* (Darwin, King, and Fitz Roy 1839).

6. The Demerson work in question is probably *La géologie enseignée en vingt-deux leçons, ou, Histoire naturelle du globe*, Paris: 1829. This work was very popular in Argentina.

7. The civil war in Argentina involved the fight between the Unitarians (centralists and supporters of European ideas) and the Federalists (supporters of Rosas,

who defended traditional colonial values such as religion and a political order of federal character).

8. Burmeister (see below) would concur with Darwin's classification of Muñiz's specimen as belonging to genus *Machaerodus*. All South American saber-tooths are now assigned to genus *Smilodon*.

9. The less than complimentary image of Argentina that some extracted from Parish's account is apparent in a review of his book published in London in 1839. In it the author disagrees with Parish's optimism about the future of the country. At least until "some enterprising population" arrived from Britain or the United States, "be scattered over such almost measureless tracts of land, and till whatever is Spanish be superseded," civilized man could not "confidently hope to see a gratifying change in either one way or another rapidly realized" ("Sir Woodbine Parish's Buenos Ayres" 1839, 515).

Chapter 2

1. Letter 11548: Coghlan, John to Darwin, C. R., 9 June 1878, www.darwinproject.ac.uk/entry-11548.

2. Hudson's findings were quickly reported in the most important scientific forums. P. L. Sclater reported in 1875, "W. H. Hudson, of Buenos Ayres has long studied the birds and other animals of that country, and deserves honourable mention in a country where so few of the native-born citizens pursue science. His bird-collections have been worked out by Mr. Salvin and myself . . . and Mr Hudson has likewise published a series of interesting notices on the habits of the species. . . . On the Rio Negro of Patagonia, where Mr. Darwin made considerable collections, we have a more recent authority in Mr. W. H. Hudson, whose series of birds from this district was examined by myself in 1871" (Sclater 1876). The works by Hudson mentioned are P. L. Sclater, M.A., and Osbert Salvin, M.A., "List of Birds collected at Conchitas, Argentine Republic, by Mr. William H. Hudson," *Proceedings of the Zoological Society*, 1868, 137; P. L. Sclater and Osbert Salvin, "Second List of Birds collected at Conchitas, Argentine Republic, by William H. Hudson, together with some notes upon another collection from the same locality," *Proceedings of the Zoological Society*, 1869, 158; P. L. Sclater and Osbert Salvin, "Third List of Birds Collected at Conchitas, Argentine Republic, by Mr. Willliam H. Hudson," *Proceedings of the Zoological Society*, 1869, 631. There were also letters from Hudson, a member of the zoological society, in 1870. Hudson's *On the Birds of the Rio Negro of Patagonia*, with notes of P. L. Sclater, was published in 1872.

3. David Brown (2006, 230) has studied the influence that Burmeister's *Black Man*, published in 1853 in the United States, had on Helper's writings. "It is possible that Helper began to see himself as the American equivalent of Burmeister and even offered his services to him." The British consul Thomas J. Hutchinson was another eager participant in these scientific exchanges. He published an essay

on the Chaco Indians in the *Revista del Plata* and shared some of his ideas with Helper. He also founded the Rosario newspaper *The Argentine Citizen* in 1864 with the intention of promoting immigration and industrial progress. Hutchinson also wrote an important book on Argentina (Hutchinson 1865).

4. For a detailed analysis of the function of the Museum see Andermann 2007.

5. Elvira Inés Baffi and María F. Torres describe him as "a rigid biblical creationist and a disciple of Georges Cuvier" (1997, 235).

6. Michelet had influenced Argentine intellectuals before this time. Domingo Sarmiento discussed his importance in a 1843 speech, also mentioning Guizot, Thierry, Niebuhr, Thiers, Cuvier, Villemain, Hugo, Chateaubriand, and Lamartime (1885a, 3:5–6).

7. An incomplete Spanish translation of Royer's French appeared in Madrid in 1872. "The first authorized Spanish translations of *Origin* and *Descent* were: Darwin, Charles. *Origen de las especies por medio de la selección natural ó la conservación de las razas favorecidas en la lucha por la existencia*. Trans. Enrique Godínez. Madrid: Biblioteca Perojo, 1877. Darwin, Charles. *La descendencia del hombre y la selección en relación al sexo*. Trans. José del Perojo and Enrique Camps. Madrid: Administración de la Revista de Medicina y Cirugía Prácticas (Rivadeneyra), 1885. An unauthorized anonymous translation predated these, which gave the first eight chapters of *The Descent* in a more or less standard translation and then summarized the second half in a section of appendices. See: Darwin, Charles. *El origen del hombre: la selección natural y la sexual (primera versión española)*. Barcelona: Renaixensa, 1876. This translation was later attributed to a Catalan poet named Joaquín Bartrina" (Travis Landry, personal communication). See also Gomis Blanco and Llorca 2007. We are grateful to an anonymous reviewer for correcting our chronology.

8. Letter 5512, Vogt, Carl to Darwin, C. R., 23 Apr 1867, www.darwinproject .ac.uk/entry-5512.

9. Letter 3595: Darwin, C. R. to Gray, Asa, 10–20 June [1862], www.darwinproject .ac.uk/entry-3595.

10. Letter 3721: Darwin, C. R. to Hooker, J. D., 11 Sept [1862], www.darwinproject .ac.uk/entry-3721.

11. Royer's second edition was published in 1866 by V. Masson et Fils, Guillaumin.

12. Socialists were significant in maintaining the importance of Burmeister. We have found evidence that, through the Paris *Vorwärts*, even Karl Marx may have come to know and appreciate his work (see Grandjonc 1974, 63).

13. So much so that the publication of the *Anales* eventually waned, such that after a hiatus of several years, Burmeister had to be reminded by his superiors where his responsibilities lay. See Andermann, n.d.

14. We are grateful to John van Wyhe for pointing us in the right direction in our search for Burmeister's source.

15. By 1870, Pasteur's refutation of the doctrine of spontaneous generation was generally accepted, especially in France.

16. A similar text of the final chapter of Burmeister's *History of Creation*, minus all reference to Darwin, appears in the sixth German edition of 1856 (Burmeister 1856a). In the 1850s this same German text, "Der Mensch, das jüngste Geschöpf der Erde," also appeared in serialized form in *Vorwärts*, a newspaper published by socialist exiles living in Paris. Karl Marx encountered it there (see Grandjonc 1974, 63).

17. There *were* post-Darwinian currents *within* German *Naturphilosophie*—or so Sandor Gliboff argues in his interpretation of Haeckel (see Gliboff 2008). We are grateful to an anomyous reviewer for alerting us to this work.

18. See Simon 1963 regarding the notion of "positive" evidence in German natural philosophy, as opposed to Comtian philosophy of science.

19. In his otherwise laudatory 1893 obituary, Otto Taschenberg, back in Halle, would call Burmeister's behavior with regard to fellow émigré scientists "despotic" during this period. Accounts had apparently made their way back to Germany, and they "reflect so poorly on the character of the great savant that it is better to pass over them in silence" (Taschenberg 1893, 46). Taschenberg also speculates that, had Burmeister secured a position in Berlin, rather than the relatively peripheral Halle, he would never have emigrated.

20. Born in Havana in 1850, Gassié was a lawyer and philologist, one of the founders of the Anthropological Society of Cuba. As a political activist, he served as cofounder of the Liberal Party. He died in 1878, less than a year after the article on Haeckel appeared. For further discussion of this article in the context of the broader Latin American reception of evolutionary thought, see Novoa and Levine 2009.

21. This "confusion" reading is common among those who have studied evolutionary thought during this period. See, e.g., Valejo and Miranda 2004, vol. 1.

22. Similar efforts may be observed in Uruguay, e.g., in Varela 1890, 21.

23. While in the United States, Sarmiento had read the work of Huxley and written an article in the *Boston Daily Advertiser* on the breeding of sheep in Argentina (see Palcos 1945).

24. In the 1870s, Alberdi frequently mentions Darwin as a new scientific authority with which to understand America (see, e.g., 1900a, 80–81).

25. Though reference to Darwin's geological fieldwork in Argentina continued (see, e.g., S. Estrada 1872, 227).

26. "Our member in London, Dr. Walter Reid, reports having delivered the diploma of honorary membership in this body to Dr. Charles Darwin" (Sociedad Científica Argentina. 1878 174). For the complete information on the letter of invitation and Reid's response, see *Anales de la Sociedad Científica Argentina*, vol. 21 (1891): 239–40.

27. Letter 11943: C. Darwin to Weyenbergh, 18 March 1879, www.darwinproject .ac.uk/darwinletters/calendar/entry-11943.html.

Chapter 3

1. Irina Podgorny has written an excellent study of representations of the Indian population during this period (1999, 109–18).
2. Though Sarmiento claimed on several occasions late in his life to have met the *Beagle* crew (without ever stating precisely where or when), we have found no independent evidence such a meeting ever took place.
3. For Moreno's relationship with Darwin, see Moreno 1879, 141, 148, 150, 151, 172, 188, 190, 194–95, 197, 274, 279, 286, 303, 321, 322, 436, 437, 448, 452, 459.
4. There is in Sclater and Hudson's book an interesting biographical sketch of Hudson on the same page that describes his activities as a naturalist in Argentina: "My fellow-author of the present work, though English in name and origin and now resident in London, is an Argentine citizen by birth. From his early childhood he was an observer of bird-life in the province of Buenos Ayres, and continued his investigations until he left the country for England a few years ago. Besides the pampas he explored the woods and marshes along the Plata, and the range of the Sierras from Cape Corrientes on the Atlantic to the Azul and Tapalquen, and made an expedition to the Rio Negro in 1871."
5. Born in Paris, Daireaux established himself in Argentina in 1868, where he became an important landowner. He was also a writer and an inspector of the school system. He died in 1916.
6. Selected columns are reproduced in this volume without their original dates of publication, which have proved difficult to ascertain. We believe this article appeared in the early 1890s. Mansilla distinguishes, on the one hand, Darwinism from evolutionism, and on the other, natural selection from descent with modification; but both "natural selection" and "descent with modification" are Darwin's expressions.
7. No such letter has been catalogued among Darwin's correspondence.
8. Ameghino was by now associated with Darwinism not only in Argentina. His work was mentioned in most of the studies who debated the new evolutionary thinking. Ernst Haeckel, for example, used his "Contribución al conocimiento de los mamíferos de la república Argentina," published in 1889 by the journal of the Academia de Ciencias de Cordoba, *Actas*. Haeckel also mentions the *Revista Argentina de Historia Natural* (see Haeckel 1898a, 31).
9. As noted above (p. 67), when we go by Burmeister's published remarks on evolution, Ameghino's caricature seems somewhat undeserved.
10. Discussion of the technical details of Ameghino's *Filogenia* would take us far afield indeed. In contemporary terms, his methodology most closely resembles a kind of statistical phenetics. Ameghino described each fossil specimen by means of a standard series of diagnostic measurements. These measurements, in turn, served as the data for the comparative analysis by means of which his method established phylogenetic relationships; the more closely the morphological measurements of two specimens coincided, the more closely related they were likely to be.

11. It is interesting to note that the names given to new species display the relationship between science and politics. Many species named by Ameghino reveal his political alliances at the time (see Podgorny 1997).

12. Ferri's *Criminal Sociology* originally appeared (in Italian) in 1884.

13. On Darwin's controversial use of teleology, see Ruse 2003, 91.

14. On the importance of this school in comparative perspective, see de Quirós 1987, 121, 206.

15. "The rejection of soft inheritance made no real headway until Weismann, in 1883 and 1884, published his germ-track theory and proposed a complete and permanent separation of soma and germ plasm. The total rejection of any inheritance of acquired characters meant a rejection of all so-called Lamarckian, Geoffroyian, or neo-Lamarckian theories of evolution. In fact, it left only two conceivable mechanisms of evolution: saltation (evolution owing to a sudden, major departing of existing norm) and selection among minor variants. Weismann adopted an uncompromising selectionism, a theory of evolution designated by Romanes (1896) as neo-Darwinism. It may be defined as the Darwinian theory of evolution without recourse to any kind of soft inheritance. Indeed, Weismann accepted most other components of Darwin's theory except pangenesis, now no longer needed" (Mayr 1988, 537).

16. The Monist League that Haeckel founded would prove very important in the formation of Nazi ideology. "Haeckel monist philosophy had an appeal across the political spectrum. . . . Although the Monist League included a wing of pacifists and leftists, it made an easy conversion to actively supporting Hitler. Unfortunately, Haeckel's greatest influence was on National Socialism" (Sapp 2003, 47). For more information on the role of Darwinism in the formation of American imperialism, see Hofstadter 1959.

17. Mendelian genetics was first taught formally at the University of La Plata in 1917 (see Stepan 1996, 70).

Chapter 4

1. "The term blending inheritance originally had no clear meaning. In a sense, this theory is the more abstract version of the general notion of 'blood' current in Europe before 1900. . . . The scientific concept of blending inheritance is less dramatic. We'll begin with a hypothetical example. If we imagine a white rabbit crossed with a black rabbit, on the blending theory of inheritance all the offspring would be gray. If the gray rabbits were crossed with each other, all their offspring would also be gray. In fact, so long as you crossed like with like, there should never be any chance of getting anything different. In this way, characters breed true" (Rose 2000, 34).

2. Quesada's allusion to the shape of the cranium is significant, and a commonplace in the evolutionary discourse of the time (see Bowler 2003, 88).

3. The extent to which the Afro-Argentine population really *did* disappear, and the rate of its supposed decrease, are subject to dispute. According to George Reid

Andrews, Buenos Aires's Afro-Argentine population "held fairly steady in size throughout the period from 1850 to 1870, though it dropped noticeably in percentage terms." But data from this and later periods were manipulated to make Argentina look like a whiter country. "The Europeans did come, and mixing did occur, but when it did not proceed as quickly as the elite wanted, they hurried the process up a bit by . . . statistical devices—transferal of Afro-Argentines from the *pardo-moreno* racial category to the white via the intermediate status of *trigueño*—and oft-repeated assertions that the black population had disappeared. To a certain extent those assertions reflected reality. If one uses 'disappear' in the sense of 'to become invisible,' then the Afro-Argentines had disappeared. Forming less than 1 percent of the capital's population by 1900, they were indeed a miniscule fragment of the city's population. But if one uses 'disappear' in the sense of 'to cease to exist,' it would be completely mistaken to say that the Afro-Argentines were gone" (Andrews 1980, 77).

4. An earlier version of the analysis of Damiana's case appeared in Novoa 2009a; used by permission.

5. The "stone-age" label may be found attributed to the Guayaquí (albeit with different connotations) as recently as in Clastres 1972.

6. Ernesto Quesada should not be confused with his father, Vicente, discussed in the preceding section.

7. "The verdict authorized the institution to continue to keep the bones that had been inventoried, and the ones that were not, were given back and buried in the locality of Tecka, Chubut" (Podgorny 1999, 10–11). Podgorny has written several articles on the fate of Indian remains in the *Museo de la Plata*: Podgorny 1995; 1998; Podgorny and Politis 1992.

Chapter 5

1. For a comparison of Sarmiento's and Alberdi's understanding of the significance of beauty, see Novoa 2007.

2. For a recent analysis of the relationship between the two in Holmberg's work, see Pérsico 2001.

3. For Darwin's references to Mantegazza, see, e.g., C. Darwin 2006, 1211ff.

4. See Haeckel 1902, 1:162–63.

5. See chapter 1 for a brief discussion of Rawson's 1845 dissertation, in which Lamarckian tendencies are clearly in evidence.

6. See chapter 1 for a brief discussion of Hudson's exchanges with Darwin.

Chapter 6

1. "Propounded by such prophets as Houston Stewart Chamberlain, Aryanism became virtual dogma in Germany after the Franco-Prussian war (1870–1871). Its unverifiability gave the myth a flexibility that made it easily adaptable also to England, where a belief in Anglo-Saxon superiority became the counterpart to Aryanism. This theory—that the Aryan (or Anglo-Saxon) had reached the highest level of civilization and was therefore destined, by nature and history, to gain increasing control over the world—was supported by elaborate historical monographs. . . . It need hardly be added that the definition of 'Aryan' remained elusive, beginning as a linguistic category but soon being understood to mean 'white native Northern European.' It was also easily translatable as 'Nordic,' which some of its adherents preferred" (Skidmore 1974, 51).

2. On the influence of Cabanis in Argentina, see Ingenieros 1915, 104; Ingenieros 1937; Ferreira 1917, 237.

3. On the *higienista* movement, see Rodríguez 2006; Salessi 1995.

4. The introduction to this book, along with the two chapters on psychology written by Jacques, were published separately under the title *Psicología* (Jacques 1923).

5. Comte "made of psychology on the one hand a physiological discipline, in the course of which he came to subscribe in part to the phrenological doctrine of Franz Gall, on the other hand a sociological discipline, since the individual, we might say, was determined not only by its biological heredity but also, probably in fact more so, by its social environment. In fact Comte drew the logical conclusion from this situation by adding, in the *Polity*, a seventh science to the hierarchy, at this apex, which he christened ethics but which in fact, in many respects, was much closer to psychology" (Simon 1963, 123).

6. Groussac's reference to "parasitic excrescence" is surely derived from Claparède, who studied parasitism; but the theory of degeneration to which he refers is not. He may have been influenced by Max Nordau, whose *Entartung* (Degeneration) appeared in 1892. We are grateful to an anonymous reviewer for this suggestion.

7. Menéndez published mostly in the *Anales del Círculo Médico*. One paper to which we will return later is his 1882 study, "La locura simulada."

8. This paper contains descriptions and photo plates of four brains of Native Argentines.

9. Both the second German edition and its English translation featured an effusive preface by Jakob's former supervisor at Erlangen, Adolf von Strümpell.

10. See Plotkin's assessment, cited above.

11. However, Wundt and Bunge differed dramatically on the issue of consciousness. "But as we have seen, the unconscious isn't the subconscious. . . . What's more, lacking a clearly demonstrative physiological basis for his [Wundt's] theory of the psyche's *unity of composition*, he judged the latter somewhat metaphysical, and

omitted it from his treatise on physiological psychology. And indeed, the doctrine suffers from the following major defects: 1. It fails clearly to identify an intermediate zone between unconsciousness and consciousness; and 2. It relies on a synthetic metaphysical principle, the *unity of composition*, on which all conscious acts are preceded, *a fortiori*, by subconscious factors" (C. Bunge 1927a, 106).

12. Bunge conceived of "pride" as a repressive mechanism preventing certain individuals from acknowledging the existence of their subconscious lives.

13. This similarity has been noted by Ricaurte Soler (1959, 77).

14. Here Bunge directly contradicts Ribot, for whom "Volition is not an event coming from no one knows where; it drives its roots into the depths of the unconscious and beyond the individual into the species and the race. It comes not from above, but from below: it is the sublimation of the lower instincts" (Ribot 1884, 150).

15. The use of an instinct theory was not a novelty. "By 1917 it was difficult to find any psychologist who questioned the instinct theory. By 1922 it was almost impossible to identify more than a handful of psychologists who still accepted the human instinct theory as a legitimate category of scientific explanation" (Cravens 1978, 191–91).

16. Bunge takes up the same idea again later in the same essay, claiming, "freedom is a sensation, the sensation of will. Will is a sensation, the sensation of consciousness. Consciousness is a sensation-representation, the sensation-representation of our psychophysical individuality. Our psychophysical individuality is a consequence of our animal life, for each animal is a 'unum per se.' Life is a mystery" (1927d, 106).

17. According to Ricaurte Soler (1959, 179), the term *social psychology* must have been first coined by Bunge, because the next work to use this expression in its title appeared in 1908.

18. In 1903, a *porteño* newspaper reported that the "Paris General Institute of Psychology spontaneously, and of its own initiative, elected Dr. Carlos Octavio Bunge to membership in the Institute, thereby adding him to the list of *Anales* collaborators, on which may be found the names of the most illustrious contemporary thinkers." This report gives us an idea of the impact of Bunge's work (J. Bunge 1965, 226).

19. The book was released by Alcan, the most prestigious publisher of psychology, and included a preface by Auguste Dietrich, who was one of the most respected translators at the time. For example, he translated, also for Alcan, Max Nordau's *Paradoxes Psychologiques* (1896) and *Paradoxes sociologiques* (1897).

20. Bunge describes his personal crisis as a typical fall from the summit all the way down to his death. Though he never mentions it, his secret homosexuality left him vulnerable to his own concerns about degeneracy and feminization. "Who hasn't dreamt, while in the grip of fever, of reaching the top of the mountain one triumphant morning? Suddenly, the ground dissolves beneath your feet, and you

are falling, falling, drowning in the shadows of the infinite abyss.... Oh indescribable agony!... Well, I have suffered this feeling of death, but awake, in real life" (C. Bunge 1903b, 1).

21. Ingenieros himself describes his relationships with these scholars, especially Ribot, in "Amigos y maestros" (Ingenieros 1908a, 341–64).

22. Spanish philosopher José Ortega y Gasset (1883–1955) took a similar approach to "Europeanization" during the same period, indicating how widespread this attitude was within the Iberian world. See the early essays in *Personas, obras, cosas* (Ortega y Gasset 1916). We are grateful to an anonymous reviewer for alerting us to this connection.

23. Because Ingenieros was himself an immigrant, he lacked the nostalgia of men like Bunge for Creole life and is consequently much more ruthless (1908b, 274).

24. In 1898 the *Anales de la Sociedad Científica Argentina* published a Spanish translation of the scientific obituary of Müller that had appeared in *Nature* the previous year (see W. F. H. B. 1898; 1897). The English text refers to Müllerian mimicry as Müller's most celebrated discovery. In the Spanish translation *mimicry* is rendered *mimetismo*, in accordance with the usual practice. We are grateful to an anonymous reviewer for alerting us to the nuances of the terminological shift from *mimetismo* to *simulación*.

25. The Baldwin Effect received the *imprimatur* of neo-Darwinian orthodoxy with the publication of George Gaylord Simpson's "The Baldwin Effect" (Simpson 1953; see Weber and Depew 2003).

Conclusion

1. For a taste of Lester Ward's work, see *Outlines of Sociology* (1899). Pesenti graduated from law school with a thesis on criminality and civilization, which was seriously reviewed by the same journal in which he published this article. See Pesenti 1901.

2. Though further discussion would go well beyond the scope of this study, we have been struck by the analogy between the contingency of certain biological processes, as understood within a broadly Darwinian framework, and the contingency of certain physical processes, as understood within Boltzmann's statistical mechanics and statistical thermodynamics. A thorough study of the parallels between the challenges both of them posed to the philosophy of science in the twentieth century is needed.

Works Cited

Acosta, Luis Eduardo. 2006. Una historia del *Periódico Zoológico* y la primera Sociedad Zoológica Argentina (1874–1881). *Academia Nacional de Ciencias, Miscelánea*, no. 105.
Agüero, Juan Fernández de. 1940. *Principios de ideología, elemental, abstractiva y oratoria*. Buenos Aires: Imprenta López.
Alberdi, Juan B. 1856. *Organizacion política y económica de la Confederación Argentina*. Besançon: José Jacquin.
———. 1886. *Bases y puntos de partida para la organización política de la República Argentina*. In *Obras Completas*, vol. 3. Buenos Aires: La Tribuna Nacional.
———. 1899. *Ensayos póstumos*. Vol. 7. Buenos Aires: Imprenta Juan Bautista Alberdi.
———. 1900a. *Ensayos póstumos*. Vol. 12. Buenos Aires: Imprenta Juan Bautista Alberdi.
———. 1900b. Ideas. In *Escritos póstumos de J.B. Alberdi*, vol. 3. Buenos Aires: Imprenta Juan Bautista Alberdi.
———. 1900c. Impresiones de viajes. In *Escritos póstumos de J.B. Alberdi*, vol. 15. Buenos Aires: Imprenta Juan Bautista Alberdi.
———. 1916. *Peregrinación de Luz del Día ó Viaje y Aventuras de la Verdad en el Nuevo Mundo*. Buenos Aires: Casa Vacaro.
Alcorta, Diego. 1824. Disertación sobre la manía aguda. Doctoral dissertation, Facultad de Medicina, Universidad de Buenos Aires, cat. no. 22509.
Allgemeiner deutscher Schulverein zur Erhaltung des Deutschtums im Auslande. 1906. *Handbuch des Deutschtums im Auslande*. Berlin: Dietrich Reimer.
Ameghino, Florentino. 1880. *La antigüedad del hombre en el Plata*. Paris: G. Masson.
———. 1889. Visión y Realidad (Alegoría científica a propósito de "Filogenia"). *Boletín del Instituto Geográfico Argentino* 10.
———. 1891. Una rápida ojeada a la evolución filogenética de los mamíferos. *Revista Argentina de Historia Natural* 1, no. 1.

———. 1904. Bibliografía. *Revista de la Universidad de Buenos Aires*, año 1, vol. 2.
———. 1950. *Conceptos fundamentales*. Buenos Aires: W.M. Jackson.
Andermann, Jens. n.d. The Museo nacional de ciencias naturales, Buenos Aires. http://www.bbk.ac.uk/ibamuseum/texts/Andermann05.htm.
———. 2007. *The Optic of the State: Visuality and Power in Argentina and Brazil*. Pittsburgh: University of Pittsburgh Press.
Andrews, George Reid. 1980. *The Afro-Argentines of Buenos Aires, 1800–1900*. Madison: University of Wisconsin Press.
Annals of the Museum of Buenos Aires. 1890. *Nature* 42 (May–October).
Anales de la Sociedad Científica Argentina. 1940. Vols. 129–130. Buenos Aires.
Arechavaleta, J. 1894. *Anales del Museo Nacional de Montevideo*. Montevideo: Dornaleche y Reyes.
Argentina, Government of. 1886. *Diario de Sesiones de la Cámara de Diputados, año 1885*. Vol. 2. Buenos Aires: Moreno y Nuñez.
———. 1904. *Registro Nacional de la República Argentina, año 1904*. Buenos Aires: Taler Tipográfico de la Penitenciaria Nacional.
Ariew, André. 2007. Under the Influence of Malthus's Law of Population Growth: Darwin Eschews the Statistical Techniques of Adolphe Quetelet. *Studies in History and Philosophy of Biological and Biomedical Sciences* 38:1–19.
Arocena, Felipe. 2003. *William Henry Hudson. Life, Literature, and Science*. London: McFarland.
Arrilli, Bernardo González. 1963. Lucio Vicente López. *Journal of Inter-American Studies* 2.
Asociación Católica de Buenos Aires. 1885. *Diario de sesiones de la primera asamblea de católicos argentinos*. Buenos Aires: Igon Hermanos.
Avery, John. 2003. *Information Theory and Evolution* London: World Scientific.
Baertschi, B. 2005. Diderot, Cabanis and Lamarck on Psycho-Physical Causality. *History and Philosophy of the Life Sciences* 27:451–63.
Baffi, Elvira Inés, and María F. Torres. 1997. Burmeister, Karl Hermann Konrad. *History of Physical Anthropology*, ed., Frank Spencer. London: Taylor & Francis.
Bagú, Sergio. 1936. *Vida ejemplar de José Ingenieros*. Buenos Aires: Editorial Claridad.
Baldwin, James Mark. 1896. A New Factor in Evolution. *American Naturalist* 30:441–51.
Barba, Enrique M. 1977. La fundación del museo y el ambiente científico de la época. In *Obra del Centenario del Museo de la Plata*. La Planta: Universidad Nacional.
Barrancos, Dora. 1992. *Cultura, educación y trabajadores 1890–1930*. Buenos Aires: CEAL.
Barrows, Susanna. 1981. *Distorting Mirrors: Visions of the Crowd in Late Nineteenth-Century France*. New Haven, CT: Yale University Press.

Bartsch, Samuel. 1869. Die Räderthiere und ihre bei Tübingen beobachteten Arten. In *Jahreshefte des Verieins für vaterländische Naturkunde*, ed. H. v. Mohl, H. v. Fehling, O. Fraas, F. Krauss and P. Zech. Stuttgart: Ebner & Seubert.

Bastiat, Federico. 1854. Economía Política. Sofismas Económicos. *El Plata cientifico y literario* 1:99–107.

Bayertz, Kurt. 1991. Biology and Beauty: Science and Aesthetics in fin-de-siècle Germany. In *Fin de Siècle and Its Legacy*, ed. M. Teich. Cambridge: Cambridge University Press.

Beatty, John. 1985. Speaking of Species: Darwin's Strategy. In *The Darwinian Heritage*, ed. D. Kohn. Princeton, NJ: Princeton University Press.

Beaumont, M. Elie de. 1843–1844. Report on M. Alcide d'Orbigny's Memoir, entitled General Considerations on the Geology of South America. *Edinburgh New Philosophical Journal* 36 (October 1843–April 1844): 42–62.

Beer, Dame Gillian. 2000. *Darwin's Plots: Evolutionary Narrative in Darwin, George Eliot and Nineteenth-Century Fiction*. Cambridge: Cambridge University Press.

Bennett, Tony. 2004. *Pasts beyond Memory: Evolution, Museums, Colonialism*. London: Routledge.

Berg, Carlos. 1895. *Carlos German Conrado Burmeister. Resena biografica*. Buenos Aires: Anales del Museo Nacional.

Bergson, Henri. 1911. *Creative Evolution*. Trans. A. Mitchell. New York: Henry Holt.

———. 1913. *Introduction to Metaphysics* New York: Macmillan.

Bernier, Desiderio. 1900. El caballo criollo. In *El caballo argentino*, ed. M. d. A. Republica Argentina. Buenos Aires: Argos.

———. 1902. Al Sr. Director de Agricultura y Ganado Don Ronaldo Tidblom. In *Le Surra Américain ou mal de caderas*, ed. F. Sivori and E. Lecler.

Bhabha, Homi K. 1990. *Nation and Narration*. London: Routledge.

Bibliografía. 1905. *Revista de la Universidad de Buenos Aires* 4.

Bilbao, Francisco. 1862. *La América en peligro*. Buenos Aires: Imprenta Berheim.

Bindman, David. 2002. *From Ape to Apollo: Aesthetics and the Idea of Race in the Eighteenth Century*. Ithaca, NY: Cornell University Press.

Biraben, Max. 1968. *German Burmeister—su vida, su obra*. Buenos Aires: Ediciones Culturales Argentinas.

Black, Max. 1962. *Models and Metaphor*. Ithaca, NY: Cornell University Press.

Bolívar, Simón. 1950. *Obras Completas*. Havana: Editorial Lex.

Bowler, Peter. 1994. Social Metaphors in Evolutionary Biology, 1870–1930. In *Biology as Society, Society as Biology: Metaphors*, ed. S. Maasen and P. Weingart. Dordrecht: Kluwer.

———. 1996. *Life's Splendid Drama*. Chicago: University of Chicago Press.

———. 2003. *Evolution: The History of an Idea*. Berkeley: University of California Press.

Boyd, Julian P. 1958. The Megalonyx, the Megatherium, and Thomas Jefferson's

Lapse of Memory. *Proceedings of the American Philosophical Society* 102: 420–35.

Boyd, Richard. 1979. Metaphor and Theory Change: What Is "Metaphor" a Metaphor For? In *Metaphor and Thought*, ed. A. Ortony. Cambridge: Cambridge University Press.

Brackenridge, Henry Marie. 1820. *Voyage to South America, Performed by Order of the American Government in the Years 1817 and 1818.* London: T. and J. Allman.

Brantlinger, Patrick. 2003. *Dark Vanishings: Discourse on the Extinction of Primitive Races, 1800–1930.* Ithaca, NY: Cornell University Press.

Broca, Paul. 1874. Le musée Moreno, à Buenos-Ayres. *Revue d'Anthropologie*.

Broda, Philippe. 1996. Commons and Veblen: Contrasting Ideas about Evolution. In *Joseph A. Schumpeter, Historian of Economics: Perspectives on the History of Economic Thought*, ed. L. S. Moss. London: Routledge.

Brown, David. 2006. *Southern Outcast: Hinton Rowan Helper and the Impending Crisis of the South.* Baton Rouge: Louisiana University Press.

Bunge, Carlos Octavio. 1902. Principes de psychologie transcendentale. *Revista jurídica y de ciencias sociales* 19:159–245.

———. 1903a. Aristocratizarse. Una concepción sintética de la historia. *Anales de la Facultad de Derecho y Ciencias Sociales* 4:329–53.

———. 1903b. *Nuestra América.* Barcelona: Imprenta de Henrich y Cia.

———. 1903c. *Principes de psychologie individuelle et socielle.* Paris: Félix Alcan.

———. 1905. *Nuestra America (Ensayo de Psicología Social).* Buenos Aires: Valerio Abeledo.

———. 1907. Introducción General al Estudio del Derecho. Conferencia Inaugural. La Tendencia Positiva en el Derecho Contemporáneo. *Revista Jurídica y de Ciencias Sociales* 1.

———. 1910. *Nuestra Patria. Libro de lectura para la educación nacional.* 20th ed. Buenos Aires: Angel Estrada.

———. 1918. *Nuestra América: ensayo de psicología* Buenos Aires: Casa Vaccaro.

———. 1926. *Sarmiento: estudio biográfico y crítico.* Madrid: Espasa Calpe.

———. 1927a. *Las tres leyes de la actividad psíquica.* In *Obras completas de Carlos Octavio Bunge*, vol. 6. Madrid: Espasa-Calpe.

———. 1927b. *Los dominios de la psicología.* In *Obras Completas de Carlos O. Bunge*, vol. 6. Madrid: Espasa-Calpe.

———. 1927c. *Notas de psicología social.* In *Obras completas de Carlos Octavio Bunge*, vol. 6. Madrid: Espasa-Calpe.

———. 1927d. *Notas para una teoría de la subconcencia-subvoluntad, 1894–1895.* In *Obras completas de Carlos Octavio Bunge*, vol. 6. Madrid: Espasa-Calpe.

———. 1927e. Notas para una teoría del instintismo. In *Obras Completas De Carlos O. Bunge, Estudios Filosoficos.* Madrid: Espasa-Calpe.

———. 1980. *Viaje á través de la estirpe y otras narraciones.* Buenos Aires: Biblioteca de "La Nacion."

Bunge, Julia M. 1965. *Vida maravillosa*. Buenos Aires: Emecé Editores.
Burmeister, Hermann. 1834. *Beiträge zur Naturgeschichte der Rankenfüsser*. Berlin: Enslin.
———. 1835. *Handbuch der Entomologie*. Vol. 1. Berlin: G. Reimer.
———. 1836. *A Manual of Entomology*. Trans. W. E. Shuckard. London: Edward Churton.
———. 1851. *Der menschliche Fuß als Charakter der Menschheit*. Leipzig: Otto Wigand.
———. 1856a. *Geschichte der Schöpfung*. Leipzig: Otto Wigand.
———. 1856b. *Systematische Übersicht der Thiere Brasiliens*. Vol. 2. Berlin: Georg Reimer.
———. 1856c. *Zoonomische Briefe: Allgemeine Darstellung der thierischen Organisation*. Leipzig: Otto Wigand.
———. 1864–1869. Gen. Machaerodus. *Anales del museo público* 1:123–38.
———. 1867. Bericht über ein Skelet von Machaerodus, im Staats-Museum zu Buenos Aires. *Abhandlungen der Naturforschungs Gesellschaft zu Halle* 10:181–97.
———. 1868. *Der Mensch*. Leipzig: Otto Wigand.
———. 1870. *Histoire de la création*. Trans. E. Maupas. Paris: F. Savy.
———. 1875. *Physikalische Beschreibung der Argentinischen Republik*. Halle: Ed. Anton.
———. 1876. *Description physique de la Republique Argentine*. Trans. E. Daireaux. Vol. 1. Paris: F. Savy.
———. 1879. *Description physique de la Republique Argentine*. Vol. 3. Buenos Aires: Emilio Coni.
Cabred, Domingo. 1895. La salud mental. In *Memorias de la Ciudad de Buenos Aires 1893–1894*. Buenos Aires.
Cabrera, Angel. 1944. *El pensamiento vivo de Ameghino*. Buenos Aires: Losada.
Carranza, Angel J. 1865. Los Anales del Museo Público de Buenos Aires. *Revista de Buenos Aires* 3 (28).
Carranza, Neptalí. 1905. *Oratoria argentina*. Vol. 5. Buenos Aires: Sesé y Larrañaga.
Carrasco, Gabriel, and A. J. Ballesteros-Zorraquin. 1888. *La Provincia de Santa Fé. Revista de su estado actual y de los progresos realizados*. Buenos Aires: P. Coni.
Cecchetto, Sergio. 2008. *La biología contra la democracia: Eugenesia, herencia, y prejuicio en Argentina, 1880–1940*. Mar del Plata: Eudem.
Cheng, Anne Anlin. 2001. *The Melancholy of Race: Psychoanalysis, Assimilation, and Hidden Grief (Race and American Culture)* Oxford: Oxford University Press.
Christianson, Robert. 1837. Biographical Memoir of Edward Turner M.D. *Edinburgh New Philosophical Journal* 23 (April–October).
Clark, Linda L. 1981. Social Darwinism in France. *The Journal of Modern History* 53, no. 1.

Clastres, Pierre. 1972. *Chronique des indiens Guayaki.* Paris: Plon.
Clift, William. 1835. *Notice on the Megatherium Brought from Buenos Aires by Woodbine Parish.* London: Richard Taylor.
Coleman, William. 1965. Cell, Nucleus, and Inheritance: An Historical Study. *Proceedings of the American Philosophical Society* 3.
Cook, David Paul. 1994. *Darwinism, War and History.* Cambridge: Cambridge University Press.
Cooper, Clayton Sedgewick. 1918. *Understanding South America.* New York: George H. Doran.
Cortina, José Antonio. 1878. Miscelanea. *Revista de Cuba* 4:441–48.
Cousin, Victor. 1828. *Cours de philosophie.* Paris: Pichon et Didier.
Cravens, Hamilton. 1978. *The Triumph of Evolution* Philadelphia: University of Pennsylvania Press.
Cronin, Helena. 1993. *The Ant and the Peacock.* Cambridge: Cambridge University Press.
Daireaux, Émile. 1877. Les dernières explorations dans La Pampa et La Patagonie. *Revue des Deux Mondes* 47:849–81.
———. 1888. *Vida y Costumbres en el Plata.* Vol. 1. Buenos Aires: Félix Lajouane.
Daireaux, Godofredo. 1887. *La cría del ganado en la Pampa.* Buenos Aires: F. Lajouane.
Darwin, Charles. 1846. *Journal of Researches into the Natural History and Geology of the Countries Visited during the Voyage of the H.M.S. Beagle.* New York: Harper and Brothers.
———. 1854. *A Monograph on the Sub-class Cirripedia, with Figures of all the Species.* London: Ray Society.
———. 1859. *On the Origin of Species.* London: John Murray.
———. 1868. *The Variation of Animals and Plants under Domestication.* 2 vols. London: John Murray.
———. 1870. Note on the Habits of the Pampas Woodpecker (*Colaptes campestris*). *Proceedings of the Zoological Society of London* 47:705–6.
———. 1874. *The Descent of Man, and Selection in Relation to Sex.* 2nd ed. London: John Murray.
———. 1876. *El origen del hombre: la selección natural y la sexual (primera versión española).* Barcelona: Renaixensa.
———. 1877. *Origen de las especies por medio de la selección natural ó la conservación de las razas favorecidas en la lucha por la existencia.* Trans. E. Godínez. Madrid: Biblioteca Perojo.
———. 1885. *La descendencia del hombre y la selección en relación al sexo.* Trans. J. d. P. a. E. Camps. Madrid: Administración de la Revista de Medicina y Cirugía Prácticas (Rivadeneyra).
———. 1985–. *The Correspondence of Charles Darwin.* Ed. F. Burkhardt et al. Cambridge: Cambridge University Press.

———. 2006 [1845–1871]. *From So Simple a Beginning: The Four Great Books of Charles Darwin*. Ed. E. O. Wilson. New York: Norton.

Darwin, Charles, P. P. King, and R. Fitz Roy. 1839. *Narrative of the Surveying Voyages of Her Majesty's Ships Adventure and Beagle between the Years 1826 and 1836. Describing Their Examination of the Southern Shores of South America and the Beagle's Circumnavigation of the Globe*. Vol. 3, *Journal and Remarks, 1832–1836*. London: Henry Colburn.

Darwin, Erasmus. 2007. *Zoonomia*. N.p.: BiblioBazaar.

Darwin Correspondence Project. http://www.darwinproject.ac.uk.

Darwin en una conferencia. 1883. In *Anuario bibliográfico de la República Argentina, 1882*, ed. A. N. Viola. Buenos Aires: M. Biedma.

Dawkins, Richard. 2006. *The Selfish Gene*. 3rd ed. Oxford: Oxford University Press.

De Angelis, Pedro. 1832. *Biografía del Señor General Arenales y juicio sobre la memoria histórica de su segunda campaña a la Sierra del Perú*. Buenos Aires: Imprenta de la Independencia.

———. 1839. Discurso preliminar al reconocimiento del Colorado. In *Diario de la navegación emprendida en 1781, desde el Rio Negro*, ed. P. de Angelis. Buenos Aires: Imprenta del Estado.

De Arenales, José. 1833. *Noticias históricas y descriptivas sobre el gran país del Chaco y Rio Bermejo, con observaciones relativas a un plan de navegacion y la colonizacion que se propone*. Buenos Aires: Hallet.

Degler, Carl N. 1991. *In Search of Human Nature: The Decline and Renewal of Darwinism in American Social Thought*. New York: Oxford University Press.

Delage, Y. 1908. Biologie Générale. *Revue Scientifique* 9:285–86.

Dellepiane, Antonio. 1907. *Estudios de filosofía jurídica y social*. Buenos Aires: Libreria Jurídica.

De Quirós, Constancio Bernaldo. 1987. *Modern Theories of Criminality*. Trans. A. d. Salvio. Littleton, CO: Fred Rothman.

The Descent of Man and Selection in Relation to Sex, by Charles Darwin. 1871. Book review. *Quarterly Journal of Science* 30 (April): 248–54.

Deutsches Meeresmuseum. 1993. *Meer und Museum*. Vol. 9. Stralsund: Deutsches Meeresmuseum.

De Vasconcellos, Federico A., A. Pereira Cabral. 1854. Aguas minerales de la costa del Uruguay. *El Plata científico y literario* I.

Dietrich, Auguste. 1903. Preface du traducteur. In *Principes de psychologie individuelle et socielle*. Paris: Félix Alcan.

Dijsktra, Bram. 1988. *Idols of Perversity*. Oxford: Oxford University Press.

Doello-Jurado, M. 1917. A Letter of Ch. Darwin in Argentina. *Nature* 99, no. 2485.

Doering, Adolfo. 1916. *Recuerdos de la expedición al Rio Negro*. Buenos Aires: La Academia.

Donghi, Tulio Halperín. 1954. Positivismo Historiográfico de José María Ramos Mejía. *Imago Mundi, Revista de Historia de la Cultura* 5.

———. 1980. Un nuevo clima de ideas. In *La Argentina del ochenta al centenario*, ed. G. Ferrari and E. Gallo. Buenos Aires: Sudamericana.

Drago, Luis María. 1888. *Hombres de presa*. 2nd ed. Buenos Aires: Félix Lajouane.

Duval, Mathias. 1886. *Le Darwinisme*. Paris: Adrien Delahaye.

Echeverría, Estéban. 1873. Revolución de Febrero en Francia. In *Obras Competas de Estéban Echeverría*, vol. 4. Buenos Aires: Imprenta de Mayo.

Edwards, Paul, ed. 1967. *Encyclopedia of Philosophy*. 8 vols. New York: Macmillan.

Ehrenreich, P. 1898. Neue Mitteilungen über die Guayaki (Steinzeitmenschen) in Paraguay. *Globus* 73.

Ellis, Havelock. 1906. *Studies in the Psychology of Sex: Sexual Selection in Man.* Philadelphia: F. A. Davis.

Ellwood, Charles. 1917. Introduction to *Criminal Sociology*, by Enrico Ferri, trans. Joseph Kelly. Boston: Little Brown.

Entomologist's Monthly Magazine, The. 1872–1873. Vol. 9.

Essig, E. O. 1936. A Sketch History of Entomology. *Osiris* 2.

Estrada, José Manuel. 1862. *El Catolicismo y la democracia. Refutación a "La América en peligro" del Señor D. Francisco Bilbao*. Buenos Aires: Imprenta Berheim.

———. [1862] 1899. El génesis de nuestra raza. In *Obras Completas de José Manuel Estrada*, ed. J. M. Garro. Buenos Aires: Pedro Igón.

———. 1905a. Congresso Católico. Discurso de clausura pronunciado el 30 de agosto de 1884. In *Discursos*. Buenos Aires: Compañia Sudamericana.

———. 1905b. El naturalismo y la educación. Conferencia dictada en el club católico el 21 de agosto de 1880. In *Discursos*. Buenos Aires: Compañia Sudamericana.

———. 1905c. La libertad y el liberalismo. Conferenica leída en la asociación católica de Buenos Aires en el año 1870. In *Discursos*. Buenos Aires: Compañia Sudamericana.

Estrada, Santiago. 1872. *Apuntes de viaje, del Plata a los Andes y del Mar Pacifico al Atlántico*. Buenos Aires: Imprenta Americana.

———. [1878] 1889. "Las Neurosis de los Hombres Célebres." In *Miscelánea*. Barcelona: Imprenta de Henrich y Cia.

Facultad de Filosofía y Letras. 1906. Sesión de 9 de mayo. *Revista de la Universidad de Buenos Aires* 5:323–24.

Falkner, Thomas. 1774. *A Description of Patagonia and the Adjoining Parts of South America*. Hereford: C. Pugh.

Feijoo, Bernardo Canal. 1955. *Constitución y Revolución: Juan Bautista Alberdi*. Buenos Aires: Fondo de Cultura Económica.

Ferreira, Alfredo J. 1917. Las Doctrinas de Cabanis y sus Proyecciones Pedagógicas. *Revista de filosofía, cultura, ciencias, educación* 3, no. 2.

Ferri, Enrico. 1900. *Socialism and Modern Science (Darwin-Spencer-Marx)*. Trans. R. Rives. La Monte, NY: International Library Publishing.
Foucault, Michel. 1965. *Madness and Civilization. A History of Insanity in the Age of Reason*. New York: Random House.
Fournier, E. 1870. Histoire de la creation. *Polybiblion: Revue bibliographique universelle* 5.
Frondizi, Risieri. 1943. Contemporary Argentine Philosophy. *Philosophy and Phenomenological Research* 4:180–86.
Galison, Peter. 1988. History, Philosophy, and the Central Metaphor. *Science in Context* 2:197–212.
Gallardo, Angel. 1904. Plan de estudio de historia natural. *Anales de la Sociedad Cientfica Argentina* 57:42–46.
———. 1908. Principios de clasificación. In *Anales de la Sociedad Científica Argentina*, vol. 46. Buenos Aires: Coni Hermanos.
Gandía, Enrique de. 1960. *Historia de las Ideas Políticas en la Argentina*. Buenos Aires: R. Depalma.
García, Juan Agustín. 1899. *Introducción al estudio de las ciencias sociales Argentinas*. Buenos Aires: Pablo Coni.
———. 1988. Ciencias sociales. In *El nacimiento de la psicología en la Argentina*, ed. H. Vezzeti. Buenos Aires: Puntosur.
Garriga, José. 1796. *Descripción del esqueletto de un quadrupedo muy corpulento y raro que se conserva en el Real Gabinete de Historia Natural de Madrid*. Madrid: Imprenta de la Viuda de don Joaquín Ibarra.
Gassié, Julian. 1877. La Antropología de Heckel [sic] y el Transformismo Unitario en Alemania. *Revista de Cuba* 2:256–63.
Gay, Claude. 1833. Aperçu sur les recherches d'histoire naturelle faites dans l'Amérique du Sud, et principalement dans le Chili, pendant les années 1830 et 1831. In *Annales des Sciences Naturelles*, vol. 28. Paris: Crochard.
Ghiodi, Delfina Varela Domínguez de. 1938. *Filosofía argentina: Los Idéologos*. Buenos Aires.
Ghirardi, Olsen A. 2000. *La Filosofia en Alberdi*. Córdoba: Academia nacional de Córdoba.
Ghiselin, Michael. 1973. Darwin and Evolutionary Psychology. *Science*, no. 4077 (March).
Girón Sierra, Alvaro. 2005. Darwinismo, Darwinismo Social e Izquierda Política (1859–1914). Reflexiones de Carácter General. In *Darwinismo social y eugenesia en el mundo latino*, ed. M. Miranda and G. Vallejo. Buenos Aires: Siglo XXI.
Gliboff, Sandor. 2008. *H.G. Bronn, Ernst Haeckel, and the Origins of German Darwinism*. Cambridge, MA: MIT Press.
Glick, Thomas F. 1989. *Darwin y el Darwinismo en el Uruguay y en América Latina*. Montevideo: Universidad de la República.
———. 2001. The Reception of Darwinism in Uruguay. In *The Reception of*

Darwinism in the Iberian World, ed. T. F. Glick, M. Á. Puig-Samper and R. Ruiz. Berlin: Springer.

Godoy, Sebastian. 1907. *Organización de la enseñanza agrícola en la Provincia de Buenos Aires*. La Plata: Taller de Impressiones Oficiales.

Goldstein, Jan. 2005. *The Post-Revolutionary Self: Politics and Psyche in France, 1750–1850*. Cambridge, MA: Harvard University Press.

Gomis Blanco, Alberto, and Jaume Josa Llorca. 2007. *Bibliografía crítica ilustrada de las obras de Darwin en España, 1857–2005*. Madrid: CSIC.

González, Ceferino. 1886. *Historia de la Filosofía*. 2nd ed. Vol. 4. Madrid: Agustín Jubera.

González Echeverría, Roberto. 1990. *Myth and Archive: A Theory of Latin American Narrative*. Cambridge: Cambridge University Press.

Gould, Stephen Jay. 2000. Church, Humboldt, and Darwin: The Tension and Harmony of Art and Science. In *Latin American Popular Culture: An Introduction*, ed. W. H. Beezley and L. A. Curcio-Nagy. Wilmington, DE: Scholarly Resources.

———. 2002. *The Structure of Evolutionary Theory*. Cambridge, MA: Harvard University Press.

Grandjonc, Jacques. 1974. *Marx et les communistes allemands à Paris*. Paris: F. Maspero.

Greene, John C. 1959. Biology and Social Theory in the Nineteenth Century: Auguste Comte and Herbert Spencer. In *Critical Problems in the History of Science*, ed. M. Claggett. Madison: University of Wisconsin Press.

Groussac, Paul. 1895. Introducción: La degeneración hereditaria. In *La locura en la historia*. Buenos Aires: Félix Lajouane.

———. 1901. *Noticia Histórica sobre la Biblioteca de Buenos Aires (1810–1901)*. Buenos Aires: Coni Hermanos.

———. 1902. Noticia biográfica del Doctor Don Diego Alcorta, y examen crítico de su obra. In *Anales de la biblioteca*, ed. P. Groussac. Buenos Aires: Coni Hermanos.

Grutzmann, Rita. 1998. *La novela naturalista en Argentina, 1880–1900*. Buenos Aires: Rodopi.

Gutiérrez, Juan María. 1835–1836. Megatherium. (Animal desconocido). *El Museo Americano o Libro de Todo el Mundo* 14.

Haeckel, Ernst. 1866. *Generelle Morphologie der Organismen*. 2nd ed. Vol. 2. Berlin: Georg Reimer.

———. 1898a. *The Last Link: Our Present Knowledge of the Descent of Man*. Trans. H. Gadow. London: Adam and Charles Back.

———. 1898b. *Über unsere gegenwärtige Kenntniss vom Ursprung des Menschen*. Bonn: Emil Strauss.

———. 1902. *Natürliche Schöpfungsgeschichte*. 10th ed. Vol. 1. Berlin: G. Reimer.

Harvey, Joy. 1997. *Almost of Man of Genius: Clémence Royer, Feminism, and Nineteenth-Century Science*. New Brunswick, NJ: Rutgers University Press.

Hayes, Carlton. 1914. The War of the Nations. *Political Science Quarterly* 29:687–707.
Helper, Hinton Rowan. 1871. *Noonday Exigencies in America*. New York: Bible Brothers.
Hirst, William Alfred. 1910. *Argentina*. New York: Charles Scribner.
Hofstadter, Richard. 1959. *Social Darwinism in American Thought*. New York: G. Braziller.
Holmberg, Eduardo Ladislao. 1882. *Carlos Roberto Darwin*. Buenos Aires: El Nacional.
Hudson, William Henry. 1870a. Letter from Mr. W.H. Hudson. *Proceedings of the Zoological Society of London* 47:112.
———. 1870b. Letter from Mr. W. H. Hudson. *Proceedings of the Zoological Society of London* 47:158–59.
———. 1895. *The Naturalist in La Plata*. 3rd ed. London: Chapman and Hall.
———. 1919. *The Book of a Naturalist*. New York: George H. Doran.
Hull, David. 1989. Charles Darwin and Nineteenth-Century Philosophies of Science. In *The Metaphysics of Evolution*. Albany: SUNY.
Humboldt, Alexander von. 1856. *Cosmos: A Sketch of a Physical Description of the Universe*. Trans. E. C. Otté. Vol. 1. New York: Harper.
Hutchinson, Thomas J. 1865. *Buenos Ayres and Argentine Gleanings*. London: Edwin Stanford.
Huxley, Leonard. 1990. *Life and Letters of Thomas Henry Huxley*. Vol. 2. New York: D. Appleton.
Huxley, Thomas Henry. 2004. Lectures and Essays. In *A Critical Examination of On the Origin of Species*: Kessinger.
Ingenieros, José. 1900. Etat actuel de nos conaissances sur l'origine de l'homme. *Anales del círculo médico argentino* 23:171–74.
———. 1901. De la simulación de la locura. *La Semana Médica* 8.
———. 1903. Psicología de los hispano-americanos. *Archivos de Psiquiatría, Criminología y Ciencias Afines* 2.
———. 1908a. Amigos y maestros. In *Al Margen de la ciencia*. Buenos Aires: Félix Lajouane.
———. 1908b. Las razas inferiores (San Vicente, 1905). In *Al Margen de la ciencia*. Buenos Aires: Félix Lajouane.
———. 1908c. On the Inferior Races. In *Al margen de la ciencia*. Buenos Aires: Félix Lajouane.
———. 1913. *Sociología argentina*. Madrid: D. Jorro.
———. 1914. Las Direcciones Filosóficas De La Cultura Argentina. www.informazion.com/delta-sistemas/marioweb/gratis/direcciones.pdf.
———. 1915. El contenido filosófico de la cultura argentina. *Revista de filosofía, cultura, ciencias, educación* 1.
———. 1917. *La simulación en la lucha por la vida*. 11th ed. Buenos Aires: L.J. Rosso.
———. 1919. *Principios de psicología*. 6th ed. Buenos Aires: L.J. Rosso.

———. 1920. *La locura en la Argentina.* Buenos Aires: Cooperativa Editorial "Buenos Aires."
———. 1937. *La evolución de las ideas en la Argentina.* Vol. 14, *Obras completas.* Buenos Aires: L.J. Rosso.
———. 1957. *Las doctrinas de Ameghino.* Ed. A. Ponce. *Obras Completas de José Ingenieros,* vol. 15. Buenos Aires: Elmer.
Jacques, Amédée. 1923. *Psicología.* Buenos Aires: La Cultura Argentina.
Jacques, Amédée, Jules Simon, and Émile Saisset. 1846. *Manuel de philosophie à l'usage des collèges.* Paris: Jouber.
Jaeger, Gustav. 1871. *Lehrbuch der allgemeinen Zoologie.* Leipzig: Ernst Günther.
Jakob, Christfried. 1899. *Atlas des gesunden und kranken Nervensystems.* 2nd ed. Munich: J. Lehmann.
———. 1901. *Atlas of the Nervous System.* London: W.B. Saunders.
———. 1906. Contribution à l'étude de la morphologie des cerveaux des Indiens. *Revista del Museo de la Plata* 12:60–87.
Kaplan, Edward. 1977. *Michelet's Poetic Vision: A Romantic Philosophy of Nature, Man, and Woman.* Amherst: University of Massachusetts Press.
Katra, William H. 1996. *The Argentine Generation of 1837: Echeverría, Alberdi, Sarmiento, Mitre.* Madison, NJ: Fairleigh Dickinson University Press.
Keller, Evelyn Fox. 1996. Language and Ideology in Evolutionary Theory: Reading Cultural Norms into Natural Law. In *Feminism and Science,* ed. E. F. Keller, Longino, H. Oxford: Oxford University Press.
Kilgore, W. J. 1960. Latin American Philosophy and the Place of Alejandro Korn. *Journal of Inter-American Studies* 2:77–82.
Klappenbach, Hugo. 2006. Recepción de la psicología alemana y francesa en la temprana psicología argentina. *Mnemosine* 2:75–86.
Klappenbach, Hugo, and Pavesi, Pablo. 1994. Una historia de la psicología en Latinoamérica. *Revista Latinoamericana de Psicología* 26:445–81.
Krasner, J. 1990. A Chaos of Delight: Perception and Illusion in Darwin's Scientific Writing. *Representations* 31:118–41.
Krueger, Felix. 1908. Despedida del Profesor Krueger. *Revista de derecho, historia y letras* 30:82–86.
Kuehn, Manfred. 2001. *Kant: A Biography* Cambridge: Cambridge University Press.
Lacroix, Frédéric. 1841. *Historia de la Patagonia, Tierra de Fuego, é Islas Malvinas.* Barcelona: Imprenta del Libral Barcelones.
Laguardia, Garibaldi, and Laguardia, Cincinato. 1919. *Argentina: Legend and History.* New York: Benjamin Sanborn.
Lahille, F. 1898. Guayaquis y Anamitas. *Revista del Museo de la Plata* 8.
Lakoff, Andrew. 2005. The Simulation of Madness: Buenos Aires, 1903. *Critical Inquiry* 31:848–73.
La Nación, editores. 1882. Necrología: Darwin. *Boletín del Instituto Geográfico Argentino* 3:132–34.

Larrazábal, Felipe. 1866. *The Life of Simon Bolivar*. Vol. 1. New York: Edward O. Jenkins.
Lastarria, J. V. 1868. Las cordilleras. *Revista de Buenos Aires* 6, no. 65.
Latzina, Francisco. 1889. *Censo general de población, edificación, comercio é industrias de la ciudad de Buenos Aires*. Buenos Aires: Compañia Sud-Americana de Billetes de Banco.
Leanza, H. A. 1992. Aspectos biográficos de German Burmeister. (1807–1892). *Ameghiniana* 29, no. 4.
Leask, Nigel. 2003. Darwin's "Second Sun": Alexander von Humboldt and the Genesis of the Voyage of the Beagle. In *Literature, Science, Psychoanalysis, 1830–1970: Essays in Honor of Gillian Beer*, ed. H. Small and T. Tate. Oxford: Oxford University Press.
Le Bon, Gustave 1895. *Psychologie des foules*. Paris: Alcan.
Lehmann-Nitsche, Robert. 1908. Relevamiento antropológico de una india guayaquí. *Revista del Museo de la Plata* 15:91–98.
Levene, Ricardo. 1911. *Los Orígenes de la democracia argentina*. Buenos Aires: J. Lajouane.
Levine, Alex. 2009. Partition Epistemology and Arguments from Analogy. *Synthese* (166): 593–600.
Levine, George. 2003. And If It Be a Pretty Woman All the Better. In *Literature, Science, Psychoanalysis, 1830–1970: Essays in Honor of Gillian Beer*. Oxford: Oxford University Press.
Levins, Richard, and Richard Lewontin. 1985. *The Dialectical Biologist*. Cambridge, MA: Harvard University Press.
Lewis, Marvin A. 1996. *Afro-Argentine Discourse: Another Dimension of the Black Diaspora*. Columbia: University of Missouri Press.
Lista, Ramón. 1879. *Viaje al país de los Tehuelches, Exploraciones en la Patagonia Austral*. Buenos Aires: Martín Biedman.
Locke, David. 1992. *Science as Writing*. New Haven, CT: Yale University Press.
Lombroso, Cesare. 1876. *L'uomo delinquente in rapporto all' antropologia, alla giurisprudenza ed alle discipline cacerarie*. Milano: Hoepli.
———. 1890. Sulla diffusione della antropologia criminale. In *Luis Maria Drago I criminali nati*. Torino: Fratelli Bocca.
Lombroso, Cesare, E. Ferri, and R. Garofalo. 1886. *Polemica in difesa della Scuola Criminale positiva*. Bologna: Nicola Zanichelli.
Lopes, Maria Margaret, and Irina Podgorny. 2000. The Shaping of Latin American Museums of Natural History, 1850–1990. *Osiris*, vol. 12, *Nature and Empire: Science and the Colonial Enterprise*, 108–18.
López, Vicente F. n.d. Evocaciones históricas. In *Colección Grandes Escritores Argentinos*, vol. 23. Buenos Aires: W. M. Norton.
———. 1865. Estudios filolójicos y etnológicos. *Revista de Buenos Aires* 3, no. 31 (November).

López Piñero, José María, and Thomas F. Glick. 1993. *El Megaterio de Bru y el Presidente Jefferson*. Valencia: University of Valencia.

Loudet, Osvaldo. 1971. *Historia de la psiquiatría argentina*. Buenos Aires: Ediciones Troquel.

Lyell, Charles. 1836. Address to the Geological Society, delivered at the Anniversary, on the 19th of February, 1836. *Proceedings of the Geological Society of London* 2, no. 44.

———. 1837. The Address of the President, Charles Lyell, jun. Esq, at the Anniversary, 1837. *London and Edinburgh Philosophical Magazine and Journal of Science* 62 (May): 388–413.

Maasen, Sabine, and Peter Weingart. 2000. *Metaphors and the Dynamics of Knowledge* London: Routledge.

Macdonell, Lady Anne Lumb. 1913. *Reminiscences of Diplomatic Life*. London: Adam and Charles Black.

Mammals of La Plata, The. 1863. *Natural History Review: A Quarterly Journal of Biological Science*.

Mañé Garzón, Fernando. 1990. *Un siglo de Darwinismo*. Montevideo: Facultad de Medicina.

Mansilla, Lucio V. 1963. Los animals desconocidos. In *Entre-nos. Causeries del Jueves*. Buenos Aires: Libreria Hachette.

Mantegazza, Paolo. 1876. *Rio de la Plata e Tenerife. Viaggi e studi di Paolo Mantegazza*. Milan: G. Brigola.

———. 2008. *The Physiology of Love and Other Writings*. Trans. D. Jacobson. Ed. N. Pireddu. Toronto: University of Toronto Press.

Márquez Miranda, Fernando. 1951. *Ameghino: Una vida heroica*. Buenos Aires: Editorial Nova.

———. 1957. *Valoración actual de Ameghino*. Buenos Aires: Editorial Perrot.

Martín, Barbara Rodríguez. 2005–2006. Juan María Gutiérrez y su contribución periodística (1833–1852) a la crítica cultural hispanoamericana. Doctoral dissertation, Universidad de La Laguna, Tenerife.

Martin, Percy Alvin. 1933. Slavery and Abolition in Brazil. *The Hispanic American Historical Review 13* 13 (2): 151–96.

Martin de Moussy, Jean Antoine V. 1860. *Description géographique et statistique de la Confédération Argentine*. Vol. 1. Paris: Didot Frères.

Massachusetts Historical Society. 1835–1855. *Proceedings of the Massachusetts Historical Society* 2.

Mayr, Ernst. 1985. *The Growth of Biological Thought*. Cambridge, MA: Harvard University Press.

———. 1988. *Toward a New Philosophy of Biology*. Cambridge, MA: Harvard University Press.

Mendizábal, Horacio. 1869. Mi canción. In *Horas de meditación* Buenos Aires: Imprenta de Buenos Aires.

Menéndez, Lucio. 1879. *Consideraciones sobre la estadística de la enajenación de la Provincia de Buenos Aires*. Buenos Aires: E.R. Coni.
Mirey, Jorge. 2004. *Curso de psicología científica*. Buenos Aires: n.p.
Molina, Wenceslao. 1903. Importancia de la Zootecnia. In *Anales de la Universidad Mayor de San Marcos de Lima*. Lima: Imprenta Liberal.
Molloy, Sylvia. 1991. *At face value: Autobiographical writing in Spanish America*. Cambridge: Cambridge University Press.
———. 2001. National Parts and Unnatural Others: A Reflection on Patrimony at the Turn of the Nineteenth Century. In *Science and the Creative Imagination in Latin America*, ed. E. Fishburn and E. L. Ortiz. London: Institute for the Study of the Americas.
Monthly Record of Current Events. 1853. *Harper's New Monthly Magazine* 7 (June–November): 835–36.
Montserrat, Marcelo. 1983. La influencia italiana en la actividad científica argentina del siglo XIX. In *Los italianos en la Argentina*, ed. F. Korn, 105–23. Buenos Aires: Fundación Giovanni Agnelli.
———. 2001. The Evolutionist Mentality in Argentina: An Ideology of Progress. In *The Reception of Darwinism in the Iberian World*, ed. T. F. Glick, M. Á. Puig-Samper, and R. Ruiz. Berlin: Springer.
Moore, Gregory. 2002. *Nietzsche, Biology and Metaphor*. Cambridge: Cambridge University Press.
Moore, James R. 1979. *The Post-Darwinian Controversies*. Cambridge: Cambridge University Press.
Moreno, François P. 1874. Description des cimentières et paraderos préhistoriques de Patagonie. *Revue d'Anthropologie*.
Moreno, Francisco P. 1879. *Viaje a la Patagonia Austral. Emprendido bajo los auspicos del gobierno nacional.* 2nd ed. Buenos Aires: Imprenta La Nación.
———. 1882. *El origen del hombre sud-americano. Razas y civilizaciones de este continente*. Buenos Aires: Imprenta Pablo Coni.
———. 1890–1891. El Museo de la Plata: rápida ojeada sobre su fundación y desarollo. *Revista del Museo de la Plata* 1.
———. 1893. *Por un ideal, ojeada retrospectiva de 25 años*. La Plata: Talleres del Museo de la Plata.
Moreno, Francisco P., and Eduardo Moreno. 1942. *Reminescencias de Franciso P. Moreno*. Buenos Aires: Imprenta Franciso P. Moreno.
Moss, Laurence S., ed. 1996. *Joseph A. Schumpeter, Historian of Economics: Perspectives on the History of Economic Thought*. London: Routledge.
Muñiz, Francisco. 1845. Muñi-felis Bonaerensis. *Gaceta Mercantil*, October 9.
Muniz, Francisco Farrer [sic]. 1833. A Case of Extensive Scabby Ulcerations, Cured by Vaccination. *London Medical and Surgical Journal* 3:172–74.
Nature, editors. 1909. Inheritance. *Nature* 79, no. 2039, 105.
NavarroViola, Miguel. 1854. Prospecto. *El Plata científico y literario* 1.

Nordau, Max. 1892. *Entartung*. 2 vols. Berlin: C. Dunker.
———. 1896. *Paradoxes Psychologiques*. Paris: Alcan.
———. 1897. *Paradoxes Sociologiques*. Paris: Alcan.
Notice of Fossil Sloths. 1843. *Zoologist* 1.
Noticias y variedades científicas y literarias. 1831. *Revista y repertorio bimestre de la Isla de Cuba* 1 (3): 361–74.
Nouzielles, Gabriela. 2000. *Ficciones Somáticas, Rosario*. Argentina: Beatriz Viterbo.
Novoa, Adriana. 2007. The Dilemmas of Male Consumption in Nineteenth-Century Argentina: Fashion, Consumerism, and Darwinism in Domingo F. Sarmiento and Juan B. Alberdi. *Journal of Latin American Studies* 39:771–95.
———. 2008. From Virile to Sterile: Feminization, Masculinity and Darwinism in Late Nineteenth-Century Argentina. *Ometeca* 12:152–75.
———. 2009a. The Act or Process of Dying Out: The Importance of Darwinian Extinction in Argentine Culture. *Science in Context* 22, no. 2, 217–44.
———. 2009b. Corrected Darwinism in Spanish America: Unity and Diversity in Sarmiento, Rodó and Vasconcelos. In *Darwin and Hispanic Literature*, ed. J. Hoeg and K. Larsen. New York: Edwin Mellen.
———. 2010. Darwinism in Spanish America: Unity and Diversity in Rodó and Vasconcelos. In *Darwin in Atlantic Cultures: Evolutionary Visions of Race, Gender, and Sexuality,* ed. Jeannette Eileen Jones and Patrick B. Sharp, 237–70. New York: Routledge.
Novoa, Adriana, and Alex Levine. 2009. Darwinism. In *The Blackwell Companion to Latin American Philosophy*, ed. Susana Nuccetelli et al. Oxford: Blackwell.
———. n.d. *¡Darwinistas!* Manuscript.
Nye, Robert. 1984. *Crime, Madness, and Politics in Modern France*. Princeton, NJ: Princeton University Press.
Nyhart, Lynn. 1995. *Biology Takes Form: Animal Morphology and the German Universities, 1800–1900*. Chicago: University of Chicago Press.
Olmo, Rosa del. 1981. *América Latina y su criminología*. Buenos Aires: Siglo XXI.
Olson, Richard. 2007. *Science and Scientism in Nineteenth-Century Europe*. Champaign: University of Illinois Press.
Orione, Julio. 1987. Florentino Ameghino y la influencia de Lamarck en la paleontologya argentina del siglo XIX. *Ouipu* 4:447–71.
Orione, Julio, and Fernando Rocchi. 1986. El Darwinismo en la Argentina. In *Todo es Historia* 228: 8–28.
Ortega y Gasset, José. 1916. *Personas, obras, cosas*. Madrid: La Lectura.
Owen, Richard. 1842. *Description of the Skeleton of an Extinct Gigantic Sloth, with Observations on the Osteology, Natural Affinities, and probably Habits of the Megatheroid Quadrupeds in general*. London: Van Voorst.
Palcos, Alberto. 1940. Prólogo. In *El Dogma Socialista*. La Plata: Universidad Nacional de la Plata.
———. 1945. Darwin, Sarmiento, y Holmberg. *La Prensa*, February 25.
Pappini, Mauricio R. 1988. The Influence of Evolutionary Biology in the Early

Development of Experimental Psychology in Argentina (1891–1930). *International Journal of Comparative Psychology* 2:131–38.

Parish, Sir Woodbine. 1839. *Buenos Ayres and the Provinces of the Rio de la Plata: Their Present State, Trade, and Debt.* London: John Murray.

———. 1852a. *Buenos Ayres and the Provinces of the Rio de la Plata.* 2nd ed. London: John Murray.

———. 1852b. *Buenos Aires y las Provincias del Rio de la Plata, desde su descubrimiento y conquista por los españoles.* Trans. J. Maeso. Buenos Aires: Benito Hortelano.

———. 1853. *Buenos Aires y las Provincias del Rio de la Plata, desde su descubrimiento y conquista por los españoles.* Trans. J. Maeso. Buenos Aires: Imprenta de Mayo.

Parisi, Giuseppe. 1907. *Storia degli Italiani nell'Argentina.* Rome: Enrico Voghera.

Parodiz, Juan José. 1981. *Darwin in the New World.* Leiden: Brill.

Pavarini, Massimo. 1983. *Control y dominación: Teorías criminológicas burguesas y proyecto hegemónico.* Buenos Aires: Siglo XXI.

Paz, Fernando Martinez. 1979. *La educación argentina.* Córdoba: Universidad Nacional de Córdoba.

Pearson, Keith Ansell. 1999. *Germinal Life: The Difference and Repetition of Deleuze.* London: Routledge.

Pérsico, Adriana Rodríguez. 2001. Las reliquias del banquete' darwinista: E. Holmberg, escritor y científico. *MLN* 116 (2): 371–91.

Perú, Government of. 1879. Estadistica parlamentaria de 1878 a 1879. Lima: Imprenta del Teatro.

Pesenti, Victor A. 1901. *Influencia de la civilización sobre el movimiento de la criminalidad.* Buenos Aires: Tailhade y Roselli.

———. 1902. La sociología y las otras ciencias. *Revista jurídica y de Ciencias Sociales* 19:21–26.

Pickering, Mary. 1993. *Auguste Comte.* Cambridge: Cambridge University Press.

Piñero, Horacio. 1900. Enseñanza actual de la psicología en Europa y América. *Anales de la Universidad de Buenos Aires* 15:117–38.

———. 1904. Conclusiones del professor de psicología experimental Doctor Horacio G. Piñero. *Revista de la Universidad de Buenos Aires* 2:391–94.

Plotkin, Mariano Ben. 2007. *Freud in the Pampas.* Palo Alto: Stanford University Press.

Podgorny, Irina. 1995. De razón a facultat: ideas acerca de las funciones del Museo de la Plata en el periodo 1890–1919. *Runa* 22:89–104.

———. 1997. De la santidad laica del científico Florentino Ameghino y el espectáculo de la ciencia en la Argentina moderna. *Entrepasados: Revista de Historia* 13:37–61.

———. 1998. Uma exibição científica dos pampas (apontamentos para uma história da formação das coleções do Museo de La Plata). *Idéias* 5:173–216.

———. 1999. *Arqueología de la educación. Textos, indicios, monumentos. La imagen de los indios en el mundo escolar.* Buenos Aires: Sociedad Argentina de Antropología.

———. 2005. Bones and Devices in the Constitution of Paleontology in Argentina at the End of the Nineteenth Century. *Science in Context* 18:249–83.

Podgorny, Irina, and Gustavo Politis. 1992. ¿Qué sucedió en la historia? Los esqueletos araucanos del Museo de La Plata y la Conquista del Desierto. *Arqueología contemporánea* 3:73–79.

Posada, Adolfo. 1902. *Literatura y problemas de sociología.* Madrid: Fernando Fé.

Pratt, Mary. 1992. *Imperial Eyes: Travel Writing and Transculturation.* London: Routledge.

Proceedings of the Royal Geographic Society and Monthly Record of Geography. 1892. Vol. 14.

Pyenson, Lewis. 1978. The Incomplete Transmission of a European Image: Physics at Greater Buenos Aires and Montreal, 1890–1920. *Proceedings of the American Philosophical Society* 122, no. 2 (April): 92–114.

———. 2002. Uses of Cultural History: Karl Lamprecht in Argentina. *Proceedings of the American Philosophical Society* 146:235–55.

Pyenson, Lewis, and Susan Sheets-Pyenson. 1999. *Servants of Nature. A History of Scientific Institutions, Enterprises, and Sensibilities.* London: W.W. Norton.

Quatrefages, Armand de. 1862. *Metamorphoses de l'homme et des animaux.* Paris: J.B. Baillière et Fils.

———. 1864. *The Metamorphosis of Man and the Lower Animals.* Trans. H. Lawson. London: Robert Hardwicke.

Quesada, Ernesto. 1903. Tristezas y esperanzas: La lucha por la vida y el descanso. *Anales de la Facultad de Derecho y Ciencias Sociales* 3:257–348.

———. 1907. *Herbert Spencer y sus doctrinas sociológicas.* Buenos Aires: Talleres gráficos de la penitenciaria nacional.

Quesada, Vicente [Victor Gálvez, pseud.]. 1882. Mi Tío Blas. *Nueva revista de Buenos Aires* 2, no. 6.

———. 1883a. La raza africana en Buenos Aires. *Nueva revista de Buenos Aires* 3, no. 8.

———. 1883b. Otros tiempos, otras costumbres. *Nueva revista de Buenos Aires* 3, no. 8.

Rafferty, Edward C. 2003. *Apostle of Human Progress: Lester Frank Ward and American Political Thought, 1841–1913.* Lanham, MD: Rowman and Littlefield.

Rafter, Nicole Hahn, and Mary Gibson. 2004. Introduction to *Criminal Woman, the Prostitute, and the Normal Woman,* by Cesare Lombroso, trans. M. Gibson. Durham, NC: Duke University Press.

Ramos Mejía, José María 1895. *La locura en la historia.* Buenos Aires: Félix Lajouane.

———. 1907. *Rosas y su tiempo.* 2nd ed. Vol. 2. Buenos Aires: Félix Lajouane.

Ramsey, Matthew. 1994. Academic Medicine and Medical Industrialism: The Regulation of Secret Remedies in Nineteenth-Century France. In *French Medical Culture in the Nineteenth Century*, ed. A. La Berge and M. Feingold. Amsterdam: Rodopi.

Rawson, Guillermo. 1891a. Observaciones sobre hygiene internacional. In *Escritos y discursos del Doctor Guillermo Rawson*. Buenos Aires: Compañía Sudamericana de Billetes de Banco.

———. 1891b. Tesis Inaugural. In *Escritos y Discursos del Doctor Guillermo Rawson*. Buenos Aires: Compañía Sudamericana de Billetes de Banco.

Resnik, Michael. 1997. *Mathematics as a Science of Patterns*. Oxford: Clarendon.

Review of Colección de obras y documentos relativos a la Historia Antigua y Moderna de las Provincias del Río de la Plata, ilustrados con notas y disertaciones. Por Pedro de Angelis, Buenos Aires: 1836. 1837. *Edinburgh Review* 65, no. 131 (April): 87–109.

Ribot, Théodule. 1884. *Les maladies de la volonté*. Paris: Félix Alcan.

Richards, Robert J. 1999. Darwin's Romantic Biology: The foundation of his evolutionary ethics. In *Biology and the Foundation of Ethics*, ed. J. Maienschein and M. Ruse. Cambridge: Cambridge University Press.

———. 2008. *The Tragic Sense of Life*. Chicago: University of Chicago Press.

Rivarola, Rodolfo. 1908. La ley universitaria y el principio de la autonomía didáctica. *Revista de la Universidad de Buenos Aires* 9:209–22.

Rodriguez, Julia. 2006. *Civilizing Argentina: Science, Medicine, and the Modern State*. Chapel Hill: University of North Carolina Press.

Rojas, Ricardo. 1903. *La victoria del hombre*. Buenos Aires: Imprenta Europea de M.A. Rosas.

Romanes, George John. 1886. *Physiological Selection*. London: Taylor and Francis.

Rose, Michael. 2000. *Darwin's Spectre*. Princeton, NJ: Princeton University Press.

Ross, Gordon. 1916. *Argentina and Uruguay*. New York: Macmillan.

Royal Geographical Society. 1841. *Journal of the Royal Geographical Society* 11.

Rudé, George. 1964. *The Crowd in History, 1730–1848*. New York: Wiley.

Rudwick, M. J. S. 1998. *Georges Cuvier, Fossil Bones, and Geological Catastrophes: New Translations and Interpretations of the Primary Texts*. Chicago: University of Chicago Press.

Ruggiero, Kristin. 2001. Passion, perversity, and the Pace of Justice in Argentina at the Turn of the last Century. In *Crime and Punishment in Latin America: Law and society since late Colonial times*, ed. C. Aguirre, R. Salvatore and G. Joseph. Durham: Duke University Press.

Ruse, Michael. 2003. *Darwin and Design: Does Evolution Have a Purpose?* Cambridge, MA: Harvard University Press.

Salessi, Jorge. 1995. *Médicos maleantes y maricas: higiene, criminología y homosexualidad en la construcción de la nación argentina (Buenos Aires, 1871–1914)*. Rosario: B. Viterbo Editora.

Salomon, Noël. 1984. *Realidad, ideología y literatura en el Facundo de D. F. Sarmiento*. Amsterdam: Rodopi.
Sanchez Sosa, Juan José, and Pablo Valderrama-Iturbe. 2001. Psychology in Latin America: Historical Reactions and Perspectives. *International Journal of Psychology* 36:384–94.
Sapp, Jan. 2003. *Genesis: The Evolution of Biology*. Oxford: Oxford University Press.
Sarmiento, Domingo F. 1849. *De la Educación Popular* Santiago: Imprenta Julio Belin.
———. 1868. *Life in the Argentine Republic in the Days of the Tyrants; or, Civilization and Barbarism*. Trans. M. Mann. New York: Macmillan.
———. 1885a. Memoria Leída en la Facultad de Humanidades. El 17 de Octubre de 1843. In *Obras de Domingo F. Sarmiento*, vol. 3. Santiago de Chile: Imprenta Gutenberg.
———. 1885b. *Vida y escritos del Coronel D. Francisco J. Muñiz*. Buenos Aires: Félix Lajouane.
———. 1899. *Obras Completas*, vol. 37–38. Buenos Aires: Ed. A. Belín Sarmiento.
———. 1900a. "Ameghino," *El Nacional*, July 13, 1882. *Obras Completas*, vol. 46. Buenos Aires: Imprenta Mariana Moreno.
———. 1900b. *Conflicto de razas y armonías en América*. In *Obras Completas*, vol. 38. Buenos Aires: Imprenta Mariano Moreno.
———. 1900c. *De la inteligencia en la vida Argentina*. In *Obras Completas*. Buenos Aires: Imprenta Mariana Moreno.
———. 1900d. *Francisco Javier Múñiz*. *Obras Completas*, vol. 46. Buenos Aires: Imprenta Mariano Moreno.
———. 1900e. *Honores al ilustre sabio M. Darwin*. In *Obras Completas*, vol. 46. Buenos Aires: Imprenta Mariano Moreno.
———. 1900f. *La escuela ultra pampeana*. *Obras Completas*, vol. 48. Buenos Aires: Imprenta Mariano Moreno.
———. 1900g. *Obras*. Vol. 36. Buenos Aires: Mariano Moreno.
———. 1900h. *Obras de D. F. Sarmiento*. Vol. 41. Buenos Aires: Mariano Moreno.
———. 1900i. "Paleontología y Arqueología Prehistórica," *El Nacional*, July 11, 1882. In *Obras Completas*, vol. 46. Buenos Aires: Imprenta Mariana Moreno.
———. 1900j. Un viaje de Nueva York a Buenos Aires del 23 de Julio al 29 de Agosto de 1868. In *Obras Completas*, vol. 49. Buenos Aires: Imprenta Mariano Moreno.
———. 1900k. *Vida de Dominguito*. Vol. 45, *Obras Completas*. Buenos Aires: Imprenta Mariana Moreno.
———. 1902. *Obras*. Vol. 52. Buenos Aires: Marquez, Zaragoza y Cia.
———. 1913. *Librería de "La Facultad."* In *Obras Completas*, vol. 3. Buenos Aires.
———. 1928. *Cuatro Conferencias*. Buenos Aires: El Ateneo.

———. 1961. *Epistolario intimo. Selección, prólogo y notas de Bernardo González Arrili.* Buenos Aires: Ediciones Culturales Argentinas.
Schaub, Edward L. 1928. *Philosophy Today: Essays on Recent Developments in the field of Philosophy.* New York: Books for Libraries Press.
Sclater, Philip Lutley. 1876. *Report of the Forty-Fifth Meeting of the British Association for the Advancement of Science; Held at Bristol in August 1875.* London: John Murray.
Sclater, Philip Lutley, and William H. Hudson. 1889. *Argentine Ornithology: A Descriptive Catalogue of the Birds of the Argentine Republic.* Vol. 2. London: R. H. Porter.
Scuder, Samuel H. 1879. *Catalog of Scientific Serials, 1633–1876.* Cambridge, MA: Library of Harvard University.
Sheets-Pyenson, Susan. 1988. *Cathedrals of Science: The Development of Colonial Natural History Museums During the Late Nineteenth Century.* Montreal: McGill-Queen's University Press.
Shumway, Jeffrey M. 2005. *The Case of the Ugly Suitor and Other Histories of Love, Gender and Nation in Buenos Aires, 1776–1870.* Lincoln: University of Nebraska Press.
Sicardi, Francisco A. 1910. *Libro extraño.* Barcelona: F. Granada.
Sighele, Scipio. 1892. *La coppia criminale.* Turin: Bocca.
Silva, Jorge Zamudio. 1940. Prólogo. In *Principios de ideología, elemental, abstractiva y oratoria.* Buenos Aires: Imprenta López.
Simon, Walter Michael. 1963. *European Positivism in the Nineteenth Century.* Ithaca, NY: Cornell University Press.
Simpson, George Gaylord. 1953. The Baldwin Effect. *Evolution* 7:110–17.
———. 1984. *Discoverers of the Lost World.* New Haven, CT: Yale University Press.
———. 1985. Extinction. *Proceedings of the American Philosophical Society* 4.
Sir Woodbine Parish's Buenos Ayres. 1839. *Monthly Review* 2, no. 4, 515–24.
Skidmore, Thomas. 1974. *Black into White.* New York: Oxford University Press.
Sluga, Hans. 1993. *Heidegger's Crisis: Philosophy and Politics in Nazi Germany.* Cambridge, MA: Harvard University Press.
Sociedad Científica Argentina. 1878. Minutes of the Meeting on February 11, 1878. *Anales de la Sociedad Cientfica Argentina* 5, no. 3 (March).
Sociedad Paleontológica de Buenos Aires. 1864–1869. *Anales del museo público de Buenos Aires* 1.
Soler, Ricaurte. 1959. *El positivismo argentino.* Buenos Aires: Imprenta Nacional.
Sommer, Doris. 1991. *Foundational Fictions: The National Romances of Latin America.* Berkeley: University of California Press.
Stagnaro, Adriana Alejandra. 1993. La antropología en la comunidad científica: entre el origen del hombre y la caza de cráneos-trofeo (1870–1910). *Alteridades* 3:53–65.

Stamos, David. 1999. Darwin's Species Category Realism. *History and Philosophy of the Life Sciences* 21:137–86.
Steinen, Karl von den. 1892. Steinzeit-Indianer in Paraguay. *Globus* 67.
Stepan, Nancy Leys. 1986. Race and Gender: The Role of Analogy in Science. *Isis* 77:261–77.
———. 1996. *The Hour of Eugenics: Race, Gender, and Nation in Latin America.* Ithaca, NY: Cornell University Press.
Taiana, Cecilia. 2005. Conceptual Resistance in the Disciplines of the Mind: The Leipzig-Buenos Aires Connection at the Beginning of the 20th Century. *History of Psychology* 8:383–402.
Taine, Hippolyte. 1875–1893. *Les origines de la France contemporaine.* Paris: Hachette.
———. 1871. *On Intelligence.* Trans. T. D. Haye. London: Reeve.
Tarde, Gabriel. 1890. *Philosophie penale.* Paris Masson.
———. 1898. *Etudes de psychologie sociale.* Paris: Giard & Brière.
Taschenberg, Otto. 1893. Karl Hermann Konrad Burmeister. *Leopoldina* 29, no. 5, 43–46.
———. 1894. *Geschichte der Zoologie und der zoologischen Sammlungen an der Universität Halle, 1694–1894.* Halle: Friedrichs-Universität.
Tassy, Pascal. 1981. Lamarck and Systematics. *Systematic Zoology* 2 (June).
Taylor, John E. 1884. *The Sagacity and Morality of Plants. A Sketch of the Life and Conduct of the Vegetable Kingdom.* London: Chatto and Windus.
Teich, Mikulás, and Roy Porter. 1990. *Fin de Siècle and Its Legacy.* Cambridge: Cambridge University Press.
Terán, Oscar. 1986. *En busca de la ideología argentina.* Buenos Aires: Catálogos.
Tomalin, Ruth. 1954. *W. H. Hudson.* London: H. F. & G. Witherby.
Torino, Inocencio. 1881. Las teorías evolucionistas y la ciencia médica. In *Nueva Revista de Buenos Aires* Buenos Aires: Imprenta de Mayo.
Triarhou, Lazaros C., and Manuel del Cerro. 2006a. The Biological Psychology of José Ingenieros, some Biographical Points, and Wilhelm Ostwald's (Nobel Prize Chemistry, 1909) Introduction to the 1922 German Edition. *Eletroneurobiología* 14:115–95.
———. 2006b. An early work in biological psychology by pioneer psychiatrist, criminologist and philosopher José Ingenieros. *Biological Psychology* 72, no. 1.
———. 2006c. Semicentennial Tribute to the Ingenious Neurobiologist Christfried Jakob (1866–1956). *European Neurology* 56:178–88.
Trübner, Karl. 1898. *Minerva, Jahrbuch der gelehrten Welt.* Strasbourg: Verlag Karl Trübner.
Unamuno, Miguel. 1902. Conferencia del Doctor Miguel Unamuno sobre la obra del Dr. C.O. Bunge, titulada, "La educación." *Revista jurídica y de ciencias sociales* 19:92–105.
Urban, Sylvanus. 1835. Geological Society. *Gentleman's Magazine.* London: William Pickering, 4:634.

Valejo, Gustavo, and Marisa Miranda. 2004. Evolución y revolución: explicaciones biológicas de utopías sociales,. In *El pensamiento alternativo en la argentina del siglo XX*, ed. H. Biagini and A. Roig. Buenos Aires: Biblos.

Varela, Juan. 1890. Juicio crítico de la primera edición. In *Vocabulario rioplatense razonado*, ed. D. Granada. Montevideo: Imprenta rural.

Vasconcelos, José. 1997. *The Cosmic Race/La Raza Cósmica*. Trans. D. T. Jaén. Baltimore: Johns Hopkins University Press.

Vezzetti, Hugo. 1985. *La locura en la Argentina*. Buenos Aires: Paidós.

———, ed. 1988. *El nacimiento de la psicología en la Argentina: pensamiento psicológico y positivismo*. Buenos Aires: Puntosur.

Victor Cousin. *Les idéologues et les Ecossais: colloque international de fevrier de 1982 au Centre international d'etudes pedagogiques, Sevres*. 1985. Paris: Presses de l'Ecole normale superieure.

Virchow, Hans. 1908. Der Kopf eines Guajaki-Mädchens. *Zeitschrift für Ethnologie* 40.

Vogt, P. F. 1903. Material zur Ethnographie und Sprache der Guayakí-Indianer. *Zeitschrift für Ethnologie* 35:849–74.

Vorzimmer, Peter. 1963. Charles Darwin and Blending Inheritance. *Isis* 54: 371–90.

W. F. H. B. 1897. Fritz Müller. *Nature* 56, no. 1458 (October 7): 546–48.

———. 1898. Fritz Müller. *Anales de la Sociedad Científica Argentina* 45:5–13.

Wallace, David Rains. 2005. *Beasts of Eden*. Berkeley: University of California Press.

Ward, Henry A. 1890–1891 [1887]. Los museos argentinos. *Revista del Museo de la Plata* 1:145–51.

Ward, Lester. 1899. *Outlines of Sociology*. New York: Macmillan.

Weber, Bruce, and David Depew, eds. 2003. *The Baldwin Effect Reconsidered*. Cambridge, MA: MIT Press.

Weismann, August. 1882. *Studies in the Theory of Descent*. New York: AMS Press.

———. 1891–1892. *Essays Upon Heredity and Kindred Biological Problems*. Oxford.

———. 1893. *The Germ Plasm: A Theory of Heredity*. London.

West, John G. 2005. Darwin's Public Policy: Nineteenth-Century Science and the Rise of the American Welfare State. In *The Progressive Revolution in Politics and Political Science: Transforming the American Regime*, ed. J. A. Marini and K. Masugi. Lanham, MD: Rowman & Littlefield.

Weyenbergh, Hendrik. 1875a. Al Lector. *Periódico zoológico* 2.

———. 1875b. Asuntos de la Sociedad. *Periódico zoológico* 2.

———. 1875c. Revista y enumeración de escritos zoológicos sobre el territorio de Sud-América. *Periódico zoológico* 2.

Wheeler, C. Gilbert. 1876. Science in the Argentine Republic. *Popular Science* 9, no. 25 (April).

Wilde, José Antonio. 1961. *Buenos Aires desde 70 años atrás*. Buenos Aires: Eudeba.
Wundt, Wilhelm. 1893. *Hipnotisme et suggestion*. Paris: Félix Alcan.
———. 1896. *Grundriss der Psychologie*. Leipzig: Wilhelm Engelmann.
———. 1897. *Outlines of Psychology*. Trans. C. H. Judd. New York: G.E. Stechert.
———. 1897–1907. *Ethics: an Investigation of the Facts and Laws of the Moral Life*. Trans. J. H. Gulliver and M. F. Washburn. New York Macmillan.
———. 1907. *Lectures on Human and Animal Psychology*. Trans. J. E. Creighton and E. B. Titchener. New York: Macmillan.
———. 1912. *An Introduction to Psychology*. Trans. R. Pintner. London: G. Allen.
———. 1928. *Elements of Folk Psychology : Outlines of a Psychological History of the Development of Mankind*. Trans. E. L. Schaub. New York: Macmillian.
———. 1969. *Principles of Physiological Psychology*. Trans. E. B. Titchener. New York: Kraus Reprint.
Young, Robert. 1890. *The Success of Christian Missions: Testimonies of their Beneficent Results*. London: Hodder and Stoughton.
———. 1900. *From Cape Horn to Panama: A Narrative of Missionary Enterprise among the Neglected Races of South America, by the South American Missionary Society*. London: South American Missionary Society.
Young, Robert M. 1985. *Darwin's Metaphor: Nature's Place in Victorian Culture*. Cambridge: Cambridge University Press.
Zeballos, Estanislao. 1877. Estudio geológico sobre la Provincia de Buenos Aires. *Anales de la Sociedad Científica Argentina* 3:2–35.

Index

Academia Nacional de las Ciencias, 70, 78–81
Afro-Argentines, 123–30, 134–35, 154, 187, 245–46
Agassiz, Louis, 68, 77, 161
Agüero, Juan Fernández de, 193–94
Alberdi, Juan Bautista: as ambassador to France, 63; *Bases y puntos de partida*, 39, 164–65; education of, 39–40; and female beauty, 170, 177, 246; first encounter with Darwinism, 77, 243; *Luz de día*, 76–77; population policy, 164–65, 180; post-Darwinian shift, 122–23; pre-Darwinian evolutionism, 39–40, 197
Alcacer, Pedro, 86
Alcorta, Diego, 194–95
Altamira, Rafael, 141–43
Alvear, Carlos María de, 172
Ameghino, Florentino: on the antiquity of South American humanity, 7, 110; attention to variation, 7, 100, 109, 245; Darwinian and Lamarckian aspects, 96–98, 109–10, 244; discussed by E. Schaub, 115; disdain for Burmeister; 54, 98–99, 244; and extinction, 7, 98, 127, 144–47, 158; humble origins, 144–45, 173; and C. Jakob, 206; nervous breakdown, 142; nonvertical evolutionary theory, 7, 99–100, 104, 109, 239, 244; peripheral consciousness, 3–4, 237; and C. Royer, 61; and D. Sarmiento, 84–88, 167–69; theory of primate radiation, 1, 7, 110; *Vision y realidad*, 123, 145–47, 153
analogy: in biology, 14, 37, 90, 213, 223; and Burmeister, 69–73; Darwin's use of, 9–10, 13, 55, 61, 76, 89, 93, 119, 162–63, 170, 178, 190, 201, 208, 223, 231–32; between evolution and metamorphosis, 57–58; and evolutionary psychology, 192, 197; E. Holmberg's use of, 173–74; and W. H. Hudson, 89; and J. Ingenieros, 224; V. López's use of, 57, 60; and W. Paley, 231; and peripheral science, 2–4, 12–13, 97, 106, 115, 119, 126, 211, 232–35; M. Pizarro's use of, 131–32; J. Rodó's use of, 158; D. Sarmiento's use of, 170–71; and sexual selection, 164, 170–73, 187; science-constitutive, 5–7, 14, 159, 164, 178, 184, 212–13, 223, 220, 231–32, 237; H. Spencer's use of, 111, 186, 197, 211, 213, 235; L. Ward's use of, 230
Angelis, Pedro de, 23–25, 32, 36, 240
Apuleius, 57
Arauz, Manuel, 295
Arenales, José de, 23–24, 26
Argerich, Cosme, 31, 193–94
Aryanism, 192, 247
Avellaneda, Nicolás, 168
Azara, Félix de, 89

Bacle, Adrienne (Andrea) Macaire, 30
Bacle, César Hipólito, 30–31
Baird, Spencer Fullerton, 53
Baldwin, James Mark, 226, 249
Bastiat, Frédéric (Federico), 47
Bates, Henry, 55, 222
Beagle, 19, 24–27, 34, 44, 64, 84, 87, 244
Beer, Dame Gillian, 12–13
Bellamy, Edward, 158–59
Bennett, Tony, 105
Berg, Carlos, 64, 80
Bergson, Henri, 107–8, 217, 236–37

Bernard, Claude, 113
Bernier, Desiderio, 189–90
Bilbao, Francisco, 49
Black, Max, 5–6
Bolívar, Simón, 22, 122
Boltzmann, Ludwig, 249
Bonpland, Aimé, 20, 44, 46, 73
Bowler, Peter, 38, 120, 173–74, 196, 199, 233, 245
Boyd, Richard, 5–6
Bravard, Auguste, 63, 73, 79, 81
Broca, Paul, 4–5, 54, 88, 151
Bru, Juan-Bautista, 17
Buffon, Georges, 5, 32–33, 37, 47, 161
Bunge, Carlos Octavio: abandonment of psychology, 2–3; and degeneracy, 188, 220; evolutionary fatalism, 156; as evolutionary psychologist, 100, 204, 214–21, 235, 247–48; and German idealism, 112, 218, 225; literary pursuits, 92, 188; and neurasthenia, 142, 219, 248–49; *Nuestra Patria*, 133–34; peripheral consciousness, 3–4
Burmeister, Hermann (German): and F. Ameghino, 64–65, 98–99, 145, 244; anti-Darwinian reputation, 54, 64–69, 79, 82, 85, 98, 159, 172, 244; and C. Darwin, 64–65, 99; dispute with other European émigrés, 70, 78, 243; early career, 62–63; and E. Haeckel, 69, 71–75, 82; and W. H. Hudson, 53, 89; and the landowning class, 55–56; and K. Marx, 242–43; mentorship of F. Moreno, 4, 54, 80, 151; and peripheral science, 62–74, 243; racial views, 161, 241; renewal of Museo Público, 46, 51–55, 63–64, 99, 241; and D. Sarmiento, 77, 82; and the Sociedad Científica, 80

Cabanis, Pierre, 193, 197, 247
Cabred, Domingo, 195, 205–6
Calzadilla, Santiago, 129
Cané, Miguel, 129
Cape Verde, 134–36
Carranza, Angel J., 56–57
Caseros, Battle of, 195
catastrophism, 32–33, 59, 119–21, 124
Catriel, Cipriano, 148–49
Chambers, Robert, 10, 39
Charcot, Jean-Martin, 114, 207, 221
Charles III, king of Spain, 17
Chascomús (province of Buenos Aires), 31–32
Cheng, Anne Anlin, 154–55

Chile, 20–21, 27, 44, 49, 59, 63, 126, 167
Círculo Médico, 85–86, 166, 168–69, 171–72, 174, 178, 180, 247
Cisneros, Baltasar Hidalgo de, viceroy of the Río de la Plata, 127–28
Claparède, René-Eduouard, 201, 247
Cobden, Richard, 59
Comte, Auguste, 247; and Argentine psychology, 196–97; and E. Bellamy, 159; and C. Bunge, 215, 220; and P. Cabanis, 193; and N. de Condorcet, 41; and the Generation of 1837, 11; incompatibilities with Darwin, 12; and J. Ingenieros, 113; and A. Jacques, 196; and A. Korn, 237; and E. Littré, 75; and E. Quesada, 234; and E. Schaub, 114. *See also* positivism
Condillac, Etienne Bonnot de, 194
Condorcet, Nicolas de Caritat, marquis de, 40–41
Cope, Edward Drinker, 100
Córdoba (city and province), 69–70, 78, 80, 168, 196
Corrientes, province of, 20, 44, 205
Cousin, Victor, 195–97, 207
criminology, 101–4, 181, 196, 198, 205, 221
criollos, 127–29, 154, 178
Cuvier, Georges: and H. Burmeister, 56, 64, 242; and J. Estrada, 47; and extinction, 119–21, 124; and D. Larrañaga, 19; and J. Lastarria, 59; and V. López, 57; and *Megatherium*, 17, 27–28, 30, 119; and F. Moreno, 151; and F. Muñiz, 32–33, 37, 64; and D. Sarmiento, 242

Daireaux, Emilio, 73, 151
Daireaux, Godofredo, 91, 244
Damiana, 136–39, 236, 246
Darwin, Caroline, 19
Darwin, Charles, doctrines: common descent, 65, 67, 69, 71, 161, 168, 183; descent with modification, 95, 244; evolution by variation and selective retention, 169, 174, 184, 190, 204, 224–25, 233; natural selection, 8–10, 34, 48, 55, 60–61, 65, 89, 95–97, 106–7, 113, 119, 121, 123–27, 130–31, 134–35, 142–43, 147, 156–64, 169–72, 174–75, 178, 181, 184–85, 189, 199, 202–3, 212, 225–26, 230, 239, 244; sexual selection, 10, 44, 92, 96, 111, 123, 126, 156–91, 203, 212, 233. *See also* analogy: Darwin's use of; *Beagle*; Burmeister, Hermann

INDEX

(German): and C. Darwin; Moreno, Francisco P.: and C. Darwin; Muñiz, Francisco; Rosas, Juan Manuel de: support of C. Darwin
Darwin, Charles, fieldwork in Argentina, 19–22, 25–30. *See also* analogy: Darwin's use of; *Beagle*; Burmeister, Hermann (German): and C. Darwin; Moreno, Francisco P.: and C. Darwin; Muñiz, Francisco; Rosas, Juan Manuel de: support of C. Darwin
Darwin, Charles, works of: *Descent of Man*, 60, 76–77, 95–96, 123, 131, 156–63, 166, 170, 172, 175, 182, 198, 229, 232, 242; *Journal of Researches*, 149–50, 240; *Origin of Species*, 9–10, 12, 21, 23–24, 37, 39, 45–46, 49, 53, 56, 58–60, 62, 67–68, 87–90, 96–97, 115, 124, 157, 159–61, 169–70, 173–75, 178, 185, 231–32, 239, 246; *Variation of Plants and Animals under Domestication*, 60. *See also* analogy: Darwin's use of; *Beagle*; Burmeister, Hermann (German): and C. Darwin; Moreno, Francisco P.: and C. Darwin; Muñiz, Francisco; Rosas, Juan Manuel de: support of C. Darwin
Darwin, Erasmus, 64
Darwinism, social, 61, 103, 240
Dellepiane, Antonio, 111–12, 212–13
Demerson, J. L., 30, 240
Destutt de Tracy, Antoine Claude Louis, 193
de Vries, Hugo, 108
Dietrich, Auguste, 217, 248
Dilthey, Wilhelm, 209, 237
Doering, Adolfo, 81
Doering, Oscar, 81
Domecq, Leon, 80
d'Orbigny, Alcide, 20, 25, 28–29, 44–46, 56, 73, 79, 81
Drago, Luis María, 101, 103–4, 181
Driesch, Hans, 107–8
Dupotet, Jean Henri, 32, 36
Durkheim, Emile, 114
Duval, Mathias, 88

Ebelot, Alfred, 125
Echeverría, Esteban, 228–29, 233
Enlightenment, 5–7, 13, 18, 23, 37, 40–41, 46, 58, 73–74, 101, 114–16, 119, 122, 126, 156, 158, 167, 171, 190, 196, 228, 233

Entre Ríos, province of, 196
Esquirol, Etienne, 194, 205
Estrada, José Manuel de, 47–49, 93–95, 132, 167, 183
Estrada, Santiago, 1, 243

Falkner, Thomas, 24
Ferri, Enrico, 101–3, 221, 245
Fisher, Ronald, 109
Fitz Roy, Robert, 24–25, 41, 240
Fournier, Eugène, 66–67
Franco-Prussian War, 180, 192, 198, 209, 247
Freycinet, Louis de, 19
Frondizi, Riseri, 236

Gaceta Mercantil, 27, 30, 33, 38, 64
Gall, Franz, 196, 247
Gallardo, Angel, 107–9, 233
Gálvez, Victor. *See* Quesada, Vicente
García, Juan Agustín, 197–99
García, Martín, 195
Garriga, José, 17–18
Gassié, Julián, 74–75, 243
gauchos, 19, 56, 179
Gay, Claude, 20–21, 44
Generation of 1837, 11, 22, 30, 39, 45, 76, 120, 122–23, 164, 173, 178, 183, 196, 228–29, 240
Generation of 1880, 173–74, 178, 183
Glick, Thomas F., 55, 61, 84, 240
Glyptodon, 26, 28, 32
Goethe, Johann Wolfgang von, 57–58, 74
Gould, Benjamin, 80
Gould, Stephen Jay, 8–9, 126
Gray, Asa, 60, 242
Groussac, Paul, 2, 194, 200–204, 247
Guayaquí people, 136–39, 246
Gutiérrez, Juan María, 30–31, 45, 119, 195

Haeckel, Ernst: and F. Ameghino, 244; Argentine impact, 90–91, 95, 134, 229; and C. Bunge, 214–16, 218; and the "Family Tree of Man," 7, 123, 201; and E. Holmberg, 173–74, 180; and ideal types, 91–92, 126, 163; and J. Ingenieros, 226, 232; and Latin American thought, 74–76, 173, 243; and ontogeny, 99; as philosophical supplement to Darwin, 12–14, 75–76, 82, 91, 105–6, 114–15, 174, 185, 210; and sexual selection, 157, 160, 163, 184; and R. Virchow, 102–3. *See also* Burmeister,

Haeckel, Ernst (*cont.*)
  Hermann (German); monism;
  Naturphilosophie
Haldane, J. B. S., 109
Hegel, Georg Wilhelm Friedrich, 7, 71, 74, 112, 236
Helper, Hinton Rowan, 53, 241–42
Henslow, John Stevens, 27
Herschell, John, 69
Hobbes, Thomas, 8, 220
Holbach, Paul-Henri Thiry de, 193
Holmberg, Eduardo, 70, 78, 86, 88, 92, 156, 164, 167, 172–78, 180, 183–85, 246
Holy Inquisition, 203
Hooker, Sir Joseph Dalton, 60, 242
Hudson, William Henry, 53, 88–90, 187, 241, 244, 246
Huergo, Luis A., 79
Humboldt, Alexander von, 8–9, 19, 21–23, 41, 47, 51, 58–59, 62–64, 67, 73, 105, 151, 156, 173, 185, 210
Huxley, Thomas Henry, 3, 71, 83, 85, 95, 115, 202, 239, 243
hybridization, 122, 190, 221

idealism, 40, 75, 82, 92, 114, 116, 156, 173, 210, 219, 230, 235
Idéologues, 40, 194–95, 204, 207
Indians (native Argentines): Aymara, 57; in the Chaco, 242; described by naturalists, 24, 95, 148–52, 206, 244, 246; extermination of, 19, 124–25, 178; extinction of, 81, 83, 87–88, 123–29, 149–52, 154, 178; Kys-Huas, 57. *See also* Damiana; Guayaquí people
Ingenieros, José: and F. Ameghino, 109–10; and degeneracy, 142; and evolutionary psychology, 104, 113–14, 134, 220–27, 235; and E. Haeckel, 232; intellectual predecessors of, 194, 204, 247, 249; and C. Jakob, 206; and J. Ramos Mejía, 194, 204; and philosophy, 235–37; and positivism, 194; racism of, 134–36; and socialism, 102, 134
Isola, Mario, 80

Jacques, Amédée, 195–96, 247
Jakob, Christfried, 206, 209–10, 247
James, William, 114, 208, 237
Janet, Pierre, 207–8

Jefferson, Thomas, 18, 33, 240
Jujuy, province of, 168

Kant, Immanuel, 5–7, 74, 112, 161, 215, 217–19, 230, 236–37
King, Philip Parker, 24, 240
Korn, Alejandro, 137, 227, 235–36
Krueger, Felix, 209–12, 214, 235
Krüger, Felix. *See* Krueger, Felix
Kuhn, Thomas S., 7

Lacroix, Frédéric, 124–25
La Cruz, Luis de, 24
Lahille, F., 136–37
Lakoff, Andrew, 222
Lamarck, Jean Baptiste, 38, 47, 49, 58–59, 65, 92, 97–98, 114, 119–20, 151, 157, 223, 226
Lamarckianism, 38, 47–48, 58, 61, 76, 96, 98, 105–6, 109, 113, 120, 170, 173, 184, 199–200, 202, 213–14, 222–23, 226, 230, 233, 245–46
La Rioja, province of, 95–96
Larrañaga, Dámaso Antonio, 19
Lastarria, José V., 59
Latzina, Francisco, 92
Le Bon, Gustave, 198
Lehmann-Nitsche, Robert, 137–39, 206
Leibniz, Gottfried Wilhelm von, 47, 196
Leroux, Pierre, 39–40, 229
Levene, Ricardo, 113
Lista, Ramón, 73, 80–81
Littré, Emile, 71–75
Lombroso, Cesare, 101, 181, 196, 221
López, Lucio V., 129
López, Vicente F., 57–60, 134, 194
Lorentz, Pablo, 70, 81
Loria, Achille, 113
Lubbock, John, 80
Lugones, Leopoldo, 221
Luján (or Luxan, province of Buenos Aires), 17, 28, 31–32, 36, 144–45
Lumb, Alfred, 52–53
Lumb, Edward, 20, 32, 34–35, 52
Lund, Peter Wilhelm, 63
Lyell, Charles, 10, 27–28, 59, 145

Maeso, Justo, 42–43
Malthus, Thomas, 9, 40, 173 223, 232
manifest destiny, doctrine of, 106–7, 111

INDEX

Mann, Thomas, 155
Mansilla, Lucio, 95–97, 244
Mantegazza, Paolo, 71, 175–77, 221, 223, 246
Martin de Moussy, Jean Antoine Victor, 43–45, 51
Marx, Karl, 101–2, 144, 159, 220, 242–43
materialism, 40, 58, 75, 82, 91, 93, 104, 112, 142, 168, 183, 195, 197, 210, 218–19, 223, 235, 237
Maupas, Emile, 66
Mayr, Ernst, 14, 163, 245
Megatherium, 17–18, 27–28, 30, 32–33, 36, 119
Mendel, Gregor, 108, 200, 232, 245
Menéndez, Lucio, 205, 247
Mercante, Victor, 235
Michelet, Jules, 57–60, 198, 242
Mill, John Stewart, 69, 159, 220
Minelli, Gustavo, 47–49
Mirey, Jorge, 208
miscegenation, 122, 189, 221
Mitre, Bartolomé, 52, 63, 77, 134, 196
modern synthesis, 7, 13, 163, 204, 239
Molina, Wenceslao F., 157
monism, 90–91, 105, 185, 210, 218, 226
Montserrat, Marcelo, 47–48, 54, 80
Moreno, Francisco P.: and P. Broca, 4–5, 54–55; and H. Burmeister, 4, 73, 80; and C. Darwin, 244; and human evolution, 83–86, 88, 105, 127, 144, 147, 156; obsession with Indian remains, 83, 100, 148–54, 206, 239; peripheral consciousness, 5; and D. Sarmiento, 85–86, 88, 167–69
Morton, Samuel, 4, 161, 239
Moulinié, Jean Jacques, 60–61
Müller, Fritz, 222, 249
Muñi-felis bonaerensis (*Smilodon* sp.), 33–34, 36, 52, 64
Muñiz, Francisco: and H. Burmeister, 52, 64, 241; and C. Darwin, 20, 31–32, 34–37, 52, 64, 241; medical research, 31–32; paleontological research, 28, 31–34, 38, 46, 64, 119, 121; peripheral consciousness, 4, 51–52, 239; and D. Sarmiento, 34, 37, 55
museo americano, El, 30–31
Museo de la Plata, 4, 105, 136, 151–53, 246
Museo Público de Buenos Aires, 18, 46, 51–56, 63–64, 99
Mylodon, 32–34, 36

Nación, La, 86–87
Nacional, El, 3, 85–86, 125
ñandú, 36–37, 121
naturalismo, 94
Naturphilosophie, 64, 173, 185, 210, 243
Navarro Viola, Miguel, 45–47
Netto, Ladislao, 80
Nietzsche, Friedrich, 147, 158, 217, 219–20, 225
Nordau, Max, 221, 247–48
Nye, Robert, 198, 200, 216

Onelli, Clemente, 206
Ortega y Gasset, José, 249
Ostwald, Wilhelm, 225
Ovid, 61
Owen, Richard, 26–27, 32–36, 54, 56, 110, 151

Pampas, 3, 18–19, 25–27, 31–32, 42, 81, 85–86, 89, 121, 125, 145, 151, 174, 244
Paraguay, 17, 44
Parchappe, Narcisse, 20
Parish, Sir Woodbine, 25–33, 36, 41–44, 165, 241
Pasteur, Louis, 2, 114, 243
Patagonia, 3, 19–20, 24–25, 27, 33, 44, 46, 80–81, 85–86, 88, 99–100, 110, 125, 148–51, 153, 172, 241
peripherality, consciousness of, 143, 152, 154, 217. *See also* Ameghino, Florentino; Bunge, Carlos Octavio; Moreno, Francisco P.; Muñiz, Francisco; Rawson, Guillermo
peripheral science, 1–7, 53, 68–70, 115, 152, 159, 192, 237–38
Pesenti, Victor R., 230, 249
Pestalozzi, Johann Heinrich, 114
phylogeny. *See* Ameghino, Florentino; Haeckel, Ernst
Pinel, Philippe, 194–95, 205
Piñero, Horacio, 207–8, 210–11, 214, 216, 235
Piñero, Norberto, 103, 208
Pius IX, pope, 47
Pizarro, Manuel, 131–32, 182–83
Podgorny, Irina, 1, 7, 52, 55–56, 105, 152, 244–46
Posada, Adolfo, 231
positivism, 2, 11, 40, 75, 79, 93, 101, 103, 114, 116, 194, 196, 211, 234, 236. *See also* Comte, Auguste; Spencer, Herbert

Posse, José, 168
Pratt, Mary Louise, 5, 22
principle of historical necessity, 169, 171, 191

Quatrefages, Armand de, 61
Quesada, Ernesto, 5, 121, 127, 141–43, 147, 183–86, 209, 225, 233–34, 246
Quesada, Vicente, 127–30, 178–79, 245, 246
Quetelet, Adolphe, 204

Ramos Mejía, José María: and criminology, 103; and heredity, 216; *La locura en la historia*, 200–205; *Rosas y su tiempo*, 187–88. See also Ingenieros, José
Rawson, Guillermo: and the hygienist movement, 194–95; and military selection, 180–81; peripheral consciousness, 4; pre-Darwinian evolutionary views, 38–39, 58, 76, 233, 246; and the Sociedad Científica, 80; students of, 200, 205
Reid, Walter F., 80, 243
Reinke, Johannes, 107–8
Renan, Ernest, 114, 144
Resnik, Michael, 6
Revue d'Anthropologie, 4, 151
Ribot, Théodule-Armand, 113, 207, 210, 216, 221, 248–49
Rivadavia, Bernardino, 18, 24, 51, 193
Rivarola, Rodolfo, 225, 228, 234–35
Roca, Julio, 81
Rodó, José, 158
Rojas, Ricardo, 143–44
Romanes, George John, 202, 226, 245
romanticism, 11–12, 40, 46, 58, 62, 114, 126, 158, 163, 171, 190, 220
Rosas, Juan Manuel de: and the Generation of 1837, 30, 33, 41–45, 49, 86, 119, 122, 143, 164, 167, 195, 199, 228, 240; and historiography, 129, 187–88; and scientific exchange, 23–25, 30–33, 51; support of C. Darwin, 19–20
Royal College of Surgeons, 26–27, 32–35
Royer, Clémence, 59–61, 66, 242
Rudwick, Martin, 17

Saint-Hilaire, Augustin, 44
Saint-Hilaire, Etienne Geoffroy, 19, 37, 39–40

Saint-Simon, Claude Henri de Rouvroy, comte de, 40, 159, 229
Saint-Vincent, Jean Bory de, 30, 161
Saisset, Emile, 196
Salta, province of, 26, 41, 168
San Juan, province of, 167, 169
San Luis, province of, 168–69
San Martin, José de, 172
Santa Fe, province of, 42, 179
Sarmiento, Domingo Faustino: and J. B. Alberdi, 77, 122–23; and the extermination of the Indians, 125; and J. Michelet, 242; pre-Darwinian evolutionism, 38–39, 120; promotion of Darwinism, 3, 77–78, 84–88, 168–78, 182, 239, 244; and sexual selection, 156, 164, 166, 171, 184. See also Ameghino, Florentino; analogy; Burmeister, Hermann (German); Cuvier, Georges; Huxley, Thomas Henry; Moreno, Francisco P.; Muñiz, Francisco; Spencer, Herbert
Schaub, Edward, 114–16, 228
Schliemann, Heinrich, 169
Schopenhauer, Arthur, 218, 237
science-constitutive analogy. See analogy: science-constitutive
Sclater, Philip Lutley, 88, 241, 244
selection, natural. See under Darwin, Charles, doctrines
selection, sexual. See under Darwin, Charles, doctrines
Senet, Rodolfo, 235
Seprun, José, 235
Sicardi, Francisco, 147–48
Siewert, Max, 70
Sighele, Scipio, 198
Simon, Jules, 196
Simon, Walter Michael, 70, 243, 247
Simpson, George Gaylord, 4, 7, 54, 63, 99, 119–21, 249
slavery, 55, 128–30, 134–35, 203
Slick, San, 150
Smith, Adam, 47, 94, 147
Snider-Pelegrini, Antonio, 169
Sociedad Científica Argentina, 79–80, 83, 86, 243, 249
Sociedad de Psicología de Buenos Aires, 235
Sociedad Paleontológica de Buenos Aires, 54, 56, 65
Soler, Ricaurte, 194, 196, 248

Sommer, Doris, 122–23
Spencer, Herbert: and J. B. Alberdi, 40, 197; discrediting of, 204–6, 208, 211–15, 217, 219, 233; and evolutionary synthesis, 75–76, 95, 101–2, 111–13, 131, 134, 167–70, 174, 183, 185–86, 198, 222–23, 225, 229, 240; as system-builder, 12–14, 105–7, 111–12, 174, 220, 226, 230–35, 237
Sprengel, Kurt, 62
Stepan, Nancy Leys, 5–6, 45, 49, 193, 200, 245
Strait of Magellan, 25, 86
Strobel, Pellegrino, 80
struggle for existence, 10, 92, 100, 103, 107, 113, 125, 131–33, 142, 146, 158, 160, 171–72, 174–75, 204, 223–24, 232
struggle for life, 48, 60, 100, 113, 132, 134–36, 141–42, 151, 157, 189, 222–25
Strümpell, Adolf von, 206, 247
Sumner, William Graham, 240
survival of the fittest, 102, 202
synthetic imperative, 13, 104–16, 214, 217, 225, 228, 235

Taine, Hippolyte, 92, 114, 193, 198–99
Tarde, Gabriel, 114, 198, 223
Tehuelche people, 150
Tejedor, Carlos, 2
ten Kate, Herman, 137
Toxodon, 54
Tribuna, La, 47
Tucumán, province of, 63, 168, 196

Uniformitarianism, 10, 59
Urquiza, Justo José de, 43, 63
Uruguay, 19–20, 55, 166, 240, 243

vaca ñata, 20, 31, 24–35, 55, 91, 202
Vaccaro, Michele Angelo, 212
Vasconcelos, José, 11, 158
Veyga, Francisco de, 104, 208
Virchow, Hans, 139–40
Virchow, Rudolf, 102, 113, 139
Visca, Pedro, 80
Vogt, Carl, 60, 68, 139, 202, 242

Wallace, Alfred Russel, 8, 54, 110–11, 163, 183–84, 187
Ward, Henry A., 151–53
Ward, Lester F., 230
Weigel Muñoz, E. J., 211–12
Weismann, August, 76, 106, 199–200, 204, 214, 223, 245
Weyenbergh, Hendrik, 70, 243
Wheeler, C. Gilbert, 78–79
Wheelwright, D. Guillermo, 52
Whewell, William, 69, 159
White, Guillermo, 80
Wilde, José Antonio, 129–30
Wright, Sewall, 109
Wundt, Wilhelm, 114, 206–7, 210–14, 216, 218–19, 230, 247

Zeballos, Estanislao, 73, 79–80, 149
Zola, Emile, 92